Operator Valued
Hardy Spaces

of the
American Mathematical Society

Number 881

Operator Valued Hardy Spaces

Tao Mei

2000 *Mathematics Subject Classification.* Primary 46L52, 32C05.

Library of Congress Cataloging-in-Publication Data

Mei, Tao, 1974–
  Operator valued Hardy spaces/Tao Mei.
    p. cm. — (Memoirs of the American Mathematical Society, ISSN 0065-9266 ; no. 881)
  "July 2007, volume 188, number 881 (second of 4 numbers)."
  Includes bibliographical references.
  ISBN 978-0-8218-3980-5 (alk. paper)
  1. Hardy spaces. 2. Operator theory. I. Title.
QA331.M475   2007
515′.94—dc22   2007060755

Memoirs of the American Mathematical Society

This journal is devoted entirely to research in pure and applied mathematics.

Subscription information. The 2007 subscription begins with volume 185 and consists of six mailings, each containing one or more numbers. Subscription prices for 2007 are US$649 list, US$519 institutional member. A late charge of 10% of the subscription price will be imposed on orders received from nonmembers after January 1 of the subscription year. Subscribers outside the United States and India must pay a postage surcharge of US$38; subscribers in India must pay a postage surcharge of US$43. Expedited delivery to destinations in North America US$53; elsewhere US$130. Each number may be ordered separately; *please specify number* when ordering an individual number. For prices and titles of recently released numbers, see the New Publications sections of the *Notices of the American Mathematical Society*.

Back number information. For back issues see the *AMS Catalog of Publications*.

Subscriptions and orders should be addressed to the American Mathematical Society, P. O. Box 845904, Boston, MA 02284-5904, USA. *All orders must be accompanied by payment.* Other correspondence should be addressed to 201 Charles Street, Providence, RI 02904-2294, USA.

Copying and reprinting. Individual readers of this publication, and nonprofit libraries acting for them, are permitted to make fair use of the material, such as to copy a chapter for use in teaching or research. Permission is granted to quote brief passages from this publication in reviews, provided the customary acknowledgment of the source is given.

Republication, systematic copying, or multiple reproduction of any material in this publication is permitted only under license from the American Mathematical Society. Requests for such permission should be addressed to the Acquisitions Department, American Mathematical Society, 201 Charles Street, Providence, Rhode Island 02904-2294, USA. Requests can also be made by e-mail to reprint-permission@ams.org.

Memoirs of the American Mathematical Society is published bimonthly (each volume consisting usually of more than one number) by the American Mathematical Society at 201 Charles Street, Providence, RI 02904-2294, USA. Periodicals postage paid at Providence, RI. Postmaster: Send address changes to Memoirs, American Mathematical Society, 201 Charles Street, Providence, RI 02904-2294, USA.

© 2007 by the American Mathematical Society. All rights reserved.
Copyright of this publication reverts to the public domain 28 years
after publication. Contact the AMS for copyright status.
This publication is indexed in *Science Citation Index*®, *SciSearch*®, *Research Alert*®,
CompuMath Citation Index®, *Current Contents*®/*Physical, Chemical & Earth Sciences*.
Printed in the United States of America.

∞ The paper used in this book is acid-free and falls within the guidelines
established to ensure permanence and durability.
Visit the AMS home page at http://www.ams.org/

10 9 8 7 6 5 4 3 2 1   12 11 10 09 08 07

Contents

Introduction	1
Chapter 1. Preliminaries	5
1. The noncommutative spaces $L^p(\mathcal{M}, L_c^2(\Omega))$	5
2. Operator valued Hardy spaces	7
3. Operator valued BMO spaces	11
Chapter 2. The Duality between $\mathcal{H}^1$ and **BMO**	15
1. The bounded map from $L^\infty(L^\infty(\mathbb{R}) \otimes \mathcal{M}, L_c^2)$ to $\text{BMO}_c(\mathbb{R}, \mathcal{M})$	15
2. The duality theorem of operator valued $\mathcal{H}^1$ and BMO	21
3. The atomic decomposition of operator valued $\mathcal{H}^1$	26
Chapter 3. The Maximal Inequality	28
1. The noncommutative Hardy-Littlewood maximal inequality	28
2. The noncommutative Lebesgue differentiation theorem and non-tangential limit of Poisson integrals	32
Chapter 4. The Duality between $\mathcal{H}^p$ and BMO^q, $1 < p < 2$.	36
1. Operator valued BMO^q ($q > 2$)	36
2. The duality theorem of $\mathcal{H}^p$ and $\text{BMO}^q (1 < p < 2)$	43
3. The equivalence of $\mathcal{H}^q$ and $\text{BMO}^q (q > 2)$	47
Chapter 5. Reduction of BMO to dyadic BMO	52
1. BMO is the intersection of two dyadic BMO	52
2. The equivalence of $\mathcal{H}_{cr}^p(\mathbb{R}, \mathcal{M})$ and $L^p(L^\infty(\mathbb{R}) \otimes \mathcal{M})(1 < p < \infty)$	53
Chapter 6. Interpolation	56
1. Complex interpolation	56
2. Real interpolation	58
3. Fourier multipliers	59
Bibliography	63

Abstract

We give a systematic study of the Hardy spaces of functions with values in the noncommutative L^p-spaces associated with a semifinite von Neumann algebra $\mathcal{M}$. This is motivated by matrix valued Harmonic Analysis (operator weighted norm inequalities, operator Hilbert transform), as well as, by the recent development of noncommutative martingale inequalities. Our noncommutative Hardy spaces are defined by noncommutative Lusin integral function. It is proved in this paper, that they are equivalent to those defined by noncommutative Littlewood-Paley G-functions. The main results of this paper include:

(i) The analogue in our setting of the classical Fefferman duality theorem between $\mathcal{H}^1$ and BMO.
(ii) The atomic decomposition of our noncommutative $\mathcal{H}^1$.
(iii) The equivalence between the norms of the noncommutative Hardy spaces and of the noncommutative L^p-spaces ($1 < p < \infty$).
(iv) The noncommutative Hardy-Littlewood maximal inequality.
(v) A description of BMO as an intersection of two dyadic BMO.
(vi) The interpolation results on these Hardy spaces.

---

[1]Received by the editor December 13, 2003, and in revised form March 4, 2005.
2000 MR Subject Classification. 46L52, 32C05.
Key words Hardy space, BMO space, Hardy-Littlewood maximal function, von Neumann algebra, noncommutative L_p space, interpolation, Lusin integral.

Introduction

This paper gives a systematic study of matrix valued (and more generally, operator valued) Hardy spaces. Our motivations come from two closely related directions. The first one is matrix valued Harmonic Analysis which deals with extending results from classical Harmonic Analysis to the operator valued setting. We should emphasize that such extensions not only are interesting in themselves but also have applications to other domains such as prediction theory and rational approximation. A central subject in this direction is the study of "operator valued" Hankel operators (i.e. Hankel matrices with operator entries). As in the scalar case, this is intimately linked to operator valued weighted norm inequalities, operator valued Carleson measures, operator valued Hardy spaces.... Much research in this direction has been done, notably by F. Nazarov, S. Treil and A. Volberg; see, for instance, the recent works [8], [24], [26], [25], [29]).

The second direction which motivates this paper is noncommutative martingale theory. This theory was initiated already in the 70's. For example, I. Cuculescu ([3]) proved a noncommutative analogue of the classical Doob weak type (1,1) maximal inequality. This has immediate applications to the almost sure convergence of noncommutative martingales (see also [12], [13]). The new input into the theory is the recent development of noncommutative martingale inequalities. This has been largely influenced and inspired by operator space theory. Many inequalities in classical martingale theory have been transferred into the noncommutative setting. These include the noncommutative Burkholder-Gundy inequalities, the noncommutative Doob inequality, the noncommutative Burkholder-Rosenthal inequalities and the boundedness of the noncommutative martingale transforms (see [28], [14], [15], [16], [31]).

One common important object in the two directions above is the noncommutative analogue of the classical BMO space. Because of the noncommutativity, there are now two noncommutative BMO spaces, the column BMO and row BMO. As expected, these noncommutative BMO spaces are proved to be the duals of some noncommutative H^1 spaces. To be more precise and to go into some details, we introduce these spaces in the case of matrix valued functions. Let $\mathcal{M}_d$ be the algebra of $d \times d$ matrices with its usual trace tr. Then the column BMO space is defined by

$$\mathrm{BMO}_c(\mathbb{R}, \mathcal{M}_d) = \{\varphi : \mathbb{R} \to \mathcal{M}_d, \|\varphi\|_{\mathrm{BMO}_c} < \infty\}$$

where

$$\|\varphi\|_{\mathrm{BMO}_c} = \sup_h \left\{ \|\varphi(\cdot)h\|_{\mathrm{BMO}(l_2^d)}, h \in l_2^d, \|h\|_{l_2^d} \leq 1 \right\}.$$

Similarly, the row BMO space is

$$\mathrm{BMO}_r(\mathbb{R}, \mathcal{M}_d) = \left\{\varphi : \mathbb{R} \to \mathcal{M}_d, \|\varphi\|_{\mathrm{BMO}_r} = \|\varphi^*\|_{\mathrm{BMO}_c} < \infty\right\}.$$

We will also need the intersection of these BMO spaces, which is
$$\mathrm{BMO}_{cr}(\mathbb{R}, \mathcal{M}_d) = \mathrm{BMO}_c(\mathbb{R}, \mathcal{M}_d) \cap \mathrm{BMO}_r(\mathbb{R}, \mathcal{M}_d)$$
equipped with the norm $\|\varphi\|_{\mathrm{BMO}_{cr}} = \max\{\|\varphi\|_{\mathrm{BMO}_c}, \|\varphi\|_{\mathrm{BMO}_r}\}$. When $d = 1$, all these BMO spaces coincide with the classical BMO space which is well known to be the dual of the classical Hardy space H^1. This result can be extended to the case of $d < \infty$ very easily. Let
$$H^1(\mathbb{R}, \mathcal{S}_d^1) = \left\{ f : \mathbb{R} \to \mathcal{S}_d^1; \int \sup_{y>0} \|f(x,y)\|_{\mathcal{S}_d^1}\, dx < \infty \right\},$$
where $\mathcal{S}_d^1$ is the trace class over l_d^2, and $f(x,y)$ denotes the Poisson integral of f corresponding to the point $x + iy$. Then
$$(H^1(\mathbb{R}, \mathcal{S}_d^1))^* = \mathrm{BMO}_{cr}(\mathbb{R}, \mathcal{M}_d)$$
and
$$c_d^{-1} \|\varphi\|_{\mathrm{BMO}_{cr}(\mathbb{R}, \mathcal{M}_d)} \le \|\varphi\|_{(H^1(\mathbb{R}, \mathcal{S}_d^1))^*} \le c_d \|\varphi\|_{\mathrm{BMO}_{cr}(\mathbb{R}, \mathcal{M}_d)}.$$
Here the constant $c_d \to +\infty$ as $d \to +\infty$. Thus this duality between $H^1(\mathbb{R}, \mathcal{S}_d^1)$ and $\mathrm{BMO}_{cr}(\mathbb{R}, \mathcal{M}_d)$ fails for the infinite dimensional case. One of our goals is to find a natural predual space of BMO_{cr} with relevant constants independent of d.

In the case of noncommutative martingales, this natural dual of BMO_{cr} was already introduced by Pisier and Xu in their work on the noncommutative Burkholder-Gundy inequality. To define the right space $\mathcal{H}^1$, they considered a noncommutative analogue of the classical square function for martingales. Motivated by their work, we will introduce a new definition of H^1 for matrix valued functions by considering a noncommutative analogue of the classical Lusin integral (recall that, in the classical case, a scalar valued function is in H^1 if and only if its Lusin integral is in L^1, see [5], [32]). For a matrix valued function f, $f \in L^1((\mathbb{R}, \frac{dt}{1+t^2}), \mathcal{M}_d), 1 \le p < \infty$, let
$$\|f\|_{\mathcal{H}_c^p(\mathbb{R}, \mathcal{M}_d)}^p = tr \int_{-\infty}^{+\infty} \Big(\iint_\Gamma |\nabla f(t+x, y)|^2 dx dy \Big)^{\frac{p}{2}} dt,$$
where $\Gamma = \{(x,y) \in \mathbb{R} : |x| < y, y > 0\}$ and
$$|\nabla f|^2 = \Big(\frac{\partial f}{\partial x}\Big)^* \frac{\partial f}{\partial x} + \Big(\frac{\partial f}{\partial y}\Big)^* \frac{\partial f}{\partial y}.$$
Then we define
$$\mathcal{H}_c^p(\mathbb{R}, \mathcal{M}_d) = \left\{ f : \mathbb{R} \to \mathcal{M}_d; \|f\|_{\mathcal{H}_c^p(\mathbb{R}, \mathcal{M}_d)} < \infty \right\}.$$
Similarly, set
$$\mathcal{H}_r^p(\mathbb{R}, \mathcal{M}_d) = \left\{ f : \mathbb{R} \to \mathcal{M}_d; \|f\|_{\mathcal{H}_r^p(\mathbb{R}, \mathcal{M}_d)} = \|f^*\|_{\mathcal{H}_c^p(\mathbb{R}, \mathcal{M}_d)} < \infty \right\}.$$
Finally, if $1 \le p < 2$, we define
$$\mathcal{H}_{cr}^p(\mathbb{R}, \mathcal{M}_d) = \mathcal{H}_c^p(\mathbb{R}, \mathcal{M}_d) + \mathcal{H}_r^p(\mathbb{R}, \mathcal{M}_d)$$
equipped with the norm
$$\|f\|_{\mathcal{H}_{cr}^p(\mathbb{R}, \mathcal{M}_d)} = \inf\{\|g\|_{\mathcal{H}_c^p} + \|h\|_{\mathcal{H}_r^p} : f = g+h, g \in \mathcal{H}_c^p(\mathbb{R}, \mathcal{M}_d), h \in \mathcal{H}_r^p(\mathbb{R}, \mathcal{M}_d)\}.$$
If $p \ge 2$, let
$$\mathcal{H}_{cr}^p(\mathbb{R}, \mathcal{M}_d) = \mathcal{H}_c^p(\mathbb{R}, \mathcal{M}_d) \cap \mathcal{H}_r^p(\mathbb{R}, \mathcal{M}_d)$$

equipped with the norm

$$\|f\|_{\mathcal{H}^p_{cr}(\mathbb{R},\mathcal{M}_d)} = \max\{\|f\|_{\mathcal{H}^p_c(\mathbb{R},\mathcal{M}_d)}, \|f\|_{\mathcal{H}^p_r(\mathbb{R},\mathcal{M}_d)}\}.$$

One of our main results is the identification of $\mathrm{BMO}_c(\mathbb{R},\mathcal{M}_d)$ as the dual of $\mathcal{H}^1_c(\mathbb{R},\mathcal{M}_d)$: $(\mathcal{H}^1_c(\mathbb{R},\mathcal{M}_d))^* = \mathrm{BMO}_c(\mathbb{R},\mathcal{M}_d)$ with equivalent norms, where the relevant equivalence constants are universal. Similarly, $\mathrm{BMO}_r(\mathbb{R},\mathcal{M}_d)$ (resp. $\mathrm{BMO}_{cr}(\mathbb{R},\mathcal{M}_d)$) is the dual of $\mathcal{H}^1_r(\mathbb{R},\mathcal{M}_d)$ (resp. $\mathcal{H}^1_{cr}(\mathbb{R},\mathcal{M}_d)$). Another result is the equality $\mathcal{H}^p_{cr}(\mathbb{R},\mathcal{M}_d) = L^p(L^\infty(\mathbb{R})\otimes\mathcal{M}_d)$ with equivalent norms for all $1 < p < \infty$. This is the function space analogue of the noncommutative Burkholder-Gundy inequality in [28]. It is also closely related to the recent work ([19]) by Junge, Le Merdy and Xu on the Littlewood-Paley theory for semigroups on noncommutative L^p-spaces.

We also prove the analogue of the classical Hardy-Littlewood maximal inequality. Our approach to this inequality for functions consists in reducing it to the same inequality for dyadic martingales. It is very simple and seems new even in the scalar case. The same idea allows to write BMO as an intersection of two dyadic BMO. This latter result plays an important role in this paper. It permits to reduce many problems involving BMO (or its variant BMO^q, which is the dual of $\mathcal{H}^p$ for $1 \leq p < 2, \frac{1}{p} + \frac{1}{q} = 1$) to dyadic BMO, that is, to BMO of dyadic noncommutative martingales. For instance, this is the case of the interpolation problems on our noncommutative Hardy spaces.

All the results mentioned above remain valid for a general semifinite von Neumann algebra $\mathcal{M}$ in place of the matrix algebras.

We now explain the organization of this paper. Chapter 1 (the next one) contains preliminaries, definitions and notations used throughout the paper. There we define the two noncommutative square functions which are the noncommutative analogues of the Lusin area integral and Littlewood-Paley g-function. These square functions allow to define the corresponding noncommutative Hardy spaces $\mathcal{H}^p_c(\mathbb{R},\mathcal{M})$, where $\mathcal{M}$ is a semifinite von Neumann algebra. This chapter also contains the definition of $\mathrm{BMO}_c(\mathbb{R},\mathcal{M})$ and some elementary properties of these spaces.

The main result of Chapter 2 is the analogue in our setting of the famous Fefferman duality theorem between $\mathcal{H}^1$ and BMO. As in the classical case, this result implies an atomic decomposition for our Hardy spaces $\mathcal{H}^1_c(\mathbb{R},\mathcal{M})$ (as well as $\mathcal{H}^1_r(\mathbb{R},\mathcal{M}), \mathcal{H}^1_{cr}(\mathbb{R},\mathcal{M})$). Another consequence is the characterization of functions in $\mathrm{BMO}_c(\mathbb{R},\mathcal{M})$ (as well as $\mathrm{BMO}_r(\mathbb{R},\mathcal{M}), \mathrm{BMO}_{cr}(\mathbb{R},\mathcal{M})$) via operator valued Carleson measures.

The objective of Chapter 3 is the noncommutative Hardy-Littlewood maximal inequality. As already mentioned above, our approach to this is to reduce this inequality to the corresponding maximal inequality for dyadic martingales. To this end, we construct two "separate" increasing filtrations $\mathcal{D} = \{\mathcal{D}_n\}_{n\in\mathbb{Z}}$ and $\mathcal{D}' = \{\mathcal{D}'_n\}_{n\in\mathbb{Z}}$ of dyadic σ-algebras. One of them is just the usual dyadic filtration on $\mathbb{R}$, while the other is a kind of translation of the first. The main point is that any interval of $\mathbb{R}$ is contained in one atom of some σ-algebra of them with comparable size. This approach will be repeatedly used in the subsequent chapters. We also prove the noncommutative Poisson maximal inequality and the noncommutative Lebesgue differentiation theorem.

In Chapter 4, we define the L^p-space analogues of the BMO spaces introduced in Chapter 1, denoted by $\mathrm{BMO}^q_c(\mathbb{R},\mathcal{M}), \mathrm{BMO}^q_r(\mathbb{R},\mathcal{M}), \mathrm{BMO}^q_{cr}(\mathbb{R},\mathcal{M})$. These spaces are proved to be the duals of the respective Hardy spaces $\mathcal{H}_c^p(\mathbb{R},\mathcal{M})$, $\mathcal{H}_r^p(\mathbb{R},\mathcal{M})$, $\mathcal{H}_{cr}^p(\mathbb{R},\mathcal{M})$ for $1 < p < 2$ ($q = \frac{p}{p-1}$). The proof of this duality is also valid for $p = 1$. In that case, we recover the duality theorem in Chapter 2. However, for $1 < p < 2$, we need, in addition, the noncommutative maximal inequality from Chapter 3. This is one of the two reasons why we have decided to present these two duality theorems separately. Another is that the reader may be more familiar with the duality between H^1 and BMO and those only interested in this duality can skip the case $1 < p < 2$. It is also proved in this chapter that $\mathrm{BMO}_c^q(\mathbb{R},\mathcal{M}) = \mathcal{H}_c^q(\mathbb{R},\mathcal{M})$ with equivalent norms for all $2 < q < \infty$. The third result of Chapter 4 is the following: Regarded as a subspace of $L^p(L^\infty(\mathbb{R}) \otimes \mathcal{M}, L_c^2(\widetilde{\Gamma}))$, $\mathcal{H}_c^p(\mathbb{R},\mathcal{M})$ is complemented in $L^p(L^\infty(\mathbb{R})\otimes\mathcal{M}, L_c^2(\widetilde{\Gamma}))$ for all $1 < p < \infty$. This result is the function space analogue of the noncommutative Stein inequality in [28]. This chapter is largely inspired by the recent work of M. Junge and Q. Xu, where the above results for noncommutative martingales were obtained.

In Chapter 5, we further exploit the reduction idea introduced in Chapter 3, in order to describe $\mathrm{BMO}_c^q(\mathbb{R},\mathcal{M})$ as $\mathrm{BMO}_c^{q,\mathcal{D}}(\mathbb{R},\mathcal{M}) \cap \mathrm{BMO}_c^{q,\mathcal{D}'}(\mathbb{R},\mathcal{M})$. These two latter BMO spaces are those of dyadic noncommutative martingales. Among the consequences given in this chapter, we mention the equivalence of $L^p(L^\infty(\mathbb{R}) \otimes \mathcal{M})$ and $\mathcal{H}_{cr}^p(\mathbb{R},\mathcal{M})$ for all $1 < p < \infty$.

Chapter 6 deals with the interpolation for our Hardy spaces. As expected, these spaces behave very well with respect to the complex and real interpolations. This chapter also contains a result on Fourier multipliers.

We close this introduction by mentioning that throughout the paper the letter c will denote an absolute positive constant, which may vary from lines to lines, and c_p a positive constant depending only on p.

CHAPTER 1

Preliminaries

1. The noncommutative spaces $L^p(\mathcal{M}, L_c^2(\Omega))$

Let $\mathcal{M}$ be a von Neumann algebra equipped with a normal semifinite faithful trace τ. Let $S_\mathcal{M}^+$ be the set of all positive x in $\mathcal{M}$ such that $\tau(\text{supp } x) < \infty$, where supp x denotes the support of x, that is, the least projection $e \in \mathcal{M}$ such that $ex = x$ (or $xe = x$). Let $S_\mathcal{M}$ be the linear span of $S_\mathcal{M}^+$. We define

$$\|x\|_p = (\tau|x|^p)^{\frac{1}{p}}, \quad \forall x \in S_\mathcal{M}$$

where $|x| = (x^*x)^{\frac{1}{2}}$. One can check that $\|\cdot\|_p$ is well-defined and is a norm on $S_\mathcal{M}$ if $1 \leq p < \infty$. The completion of $(S_\mathcal{M}, \|\cdot\|_p)$ is denoted by $L^p(\mathcal{M})$ which is the usual noncommutative L^p space associated with $(\mathcal{M}, \tau)$. For convenience, we usually set $L^\infty(\mathcal{M}) = \mathcal{M}$ equipped with the operator norm $\|\cdot\|_\mathcal{M}$. The elements in $L^p(\mathcal{M}, \tau)$ can also be viewed as closed densely defined operators on H (H being the Hilbert space on which $\mathcal{M}$ acts). We refer to [4] for more information on noncommutative L^p spaces.

Let (Ω, μ) be a measurable space. We say h is a $S_\mathcal{M}$-valued simple function on (Ω, μ) if it can be written as

$$(1.1) \qquad h = \sum_{i=1}^n m_i \cdot \chi_{A_i}$$

where $m_i \in S_\mathcal{M}$ and A_i's are measurable disjoint subsets of Ω with $\mu(A_i) < \infty$. For such a function h we define

$$\|h\|_{L^p(\mathcal{M}, L_c^2(\Omega))} = \left\|\left(\sum_{i=1}^n m_i^* m_i \cdot \mu(A_i)\right)^{\frac{1}{2}}\right\|_{L^p(\mathcal{M})}$$

and

$$\|h\|_{L^p(\mathcal{M}, L_r^2(\Omega))} = \left\|\left(\sum_{i=1}^n m_i m_i^* \cdot \mu(A_i)\right)^{\frac{1}{2}}\right\|_{L^p(\mathcal{M})}$$

This gives two norms on the family of all such h's. To see that, denoting by $B(L^2(\Omega))$ the space of all bounded operators on $L^2(\Omega)$ with its usual trace tr, we consider the von Neumann algebra tensor product $\mathcal{M} \otimes B(L^2(\Omega))$ with the product trace $\tau \otimes tr$. Given a set $A_0 \subset \Omega$ with $\mu(A_0) = 1$, any element of the family of h's above can be regarded as an element in $L^p(\mathcal{M} \otimes B(L^2(\Omega)))$ via the following map:

$$(1.2) \qquad h \mapsto T(h) = \sum_{i=1}^n m_i \otimes (\chi_{A_i} \otimes \chi_{A_0})$$

and
$$\|h\|_{L^p(\mathcal{M};L_c^2(\Omega))} = \|T(h)\|_{L^p(\mathcal{M}\otimes B(L^2(\Omega)))}$$
Therefore, $\|\cdot\|_{L^p(\mathcal{M};L_c^2(\Omega))}$ defines a norm on the family of the h's. The corresponding completion (for $1 \leq p < \infty$) is a Banach space, denoted by $L^p(\mathcal{M};L_c^2(\Omega))$. Then $L^p(\mathcal{M};L_c^2(\Omega))$ is isometric to the column subspace of $L^p(\mathcal{M}\otimes B(L^2(\Omega)))$. For $p = \infty$ we let $L^\infty(\mathcal{M};L_c^2(\Omega))$ be the Banach space isometric by the above map T to the column subspace of $L^\infty(\mathcal{M}\otimes B(L^2(\Omega)))$.

Similarly to $\|\cdot\|_{L^p(\mathcal{M};L_c^2(\Omega))}$, $\|\cdot\|_{L^p(\mathcal{M};L_r^2(\Omega))}$ is also a norm on the family of $S_\mathcal{M}$-valued simple functions and it defines the Banach space $L^p(\mathcal{M};L_r^2(\Omega))$ which is isometric to the row subspace of $L^p(\mathcal{M} \otimes B(L^2(\Omega)))$.

Alternatively, we can fix an orthonormal basis of $L^2(\Omega)$. Then any element of $L^p(\mathcal{M} \otimes B(L^2(\Omega)))$ can be identified with an infinite matrix with entries in $L^p(\mathcal{M})$. Accordingly, $L^p(\mathcal{M};L_c^2(\Omega))$ (resp. $L^p(\mathcal{M};L_r^2(\Omega))$) can be identified with the subspace of $L^p(\mathcal{M}\otimes B(L^2(\Omega)))$ consisting of matrices whose entries are all zero except those in the first column (resp. row).

PROPOSITION 1.1. *Let* $f \in L^p(\mathcal{M};L_c^2(\Omega)), g \in L^q(\mathcal{M};L_c^2(\Omega))(1 \leq p,q \leq \infty)$, $\frac{1}{r} = \frac{1}{p} + \frac{1}{q}$. *Then* $\langle g, f \rangle$ *exists as an element in* $L^r(\mathcal{M})$ *and*
$$\|\langle g, f\rangle\|_{L^r(\mathcal{M})} \leq \|g\|_{L^q(\mathcal{M};L_c^2(\Omega))} \|f\|_{L^p(\mathcal{M};L_c^2(\Omega))},$$
where $\langle\,,\,\rangle$ *denotes the scalar product in* $L_c^2(\Omega)$. *A similar statement also holds for row spaces.*

Proof. This is clear from the discussion above via the matrix representation of $L^p(\mathcal{M};L_c^2(\Omega))$ (in an orthonormal basis of $L^2(\Omega)$). ∎

Remark. Note that if f and g are $S_\mathcal{M}$-valued simple functions, then
$$\langle g, f\rangle = \int_\Omega g^* f d\mu.$$
For general f and g as in Proposition 1.1, if one of p and q is finite, one can easily prove that $\langle g, f\rangle$ is the limit in $L^r(\mathcal{M})$ of a sequence $(\langle g_n, f_n\rangle)_n$ with $S_\mathcal{M}$-valued simple functions f_n, g_n. Consequently, we can define $\int_\Omega g^* f d\mu$ as the limit of $\int_\Omega g_n^* f_n d\mu$. If both p and q are infinite, this limit procedure is still valid but only in the w*-sense.

Convention. Throughout this paper whenever we are in the situation of Proposition 1.1, we will write $\langle g, f\rangle$ as the integral $\int_\Omega g^* f d\mu$. Notationally, this is clearer. Moreover, by the proceding remark this indeed makes sense in many cases.

Observe that the column and row subspaces of $L^p(\mathcal{M} \otimes B(L^2(\Omega)))$ are 1-complemented subspaces. Therefore, from the classical duality between $L^p(\mathcal{M} \otimes B(L^2(\Omega)))$ and $L^q(\mathcal{M} \otimes B(L^2(\Omega)))$ ($\frac{1}{p} + \frac{1}{q} = 1, 1 \leq p < \infty$) we deduce that
$$\left(L^p(\mathcal{M};L_c^2(\Omega))\right)^* = L^q(\mathcal{M};L_c^2(\Omega))$$
and
$$\left(L^p(\mathcal{M};L_r^2(\Omega))\right)^* = L^q(\mathcal{M};L_r^2(\Omega))$$
isometrically via the antiduality
$$(f, g) \mapsto \tau(\langle g, f\rangle) = \tau \int_\Omega g^* f d\mu.$$

Moreover, it is well known that (by the same reason), for $0 < \theta < 1$ and $1 \leq p_0, p_1, p_\theta \leq \infty$ with $\frac{1}{p_\theta} = \frac{1-\theta}{p_0} + \frac{\theta}{p_1}$, we have isometrically

$$(1.3) \qquad \left(L^{p_0}(\mathcal{M}; L_c^2(\Omega)), L^{p_1}(\mathcal{M}; L_c^2(\Omega))\right)_\theta = L^{p_\theta}(\mathcal{M}; L_c^2(\Omega)).$$

In the following, we are mainly interested in the spaces $L^p(\mathcal{M}; L_c^2(\Omega))$ (resp. $L^p(\mathcal{M}; L_r^2(\Omega))$) with $(\Omega, \mu) = \widetilde{\Gamma} = (\Gamma, dxdy) \times (\{1,2\}, \sigma)$, where $\Gamma = \{(x,y) \in \mathbb{R}_+^2, |x| < y\}$, $\sigma\{1\} = \sigma\{2\} = 1$. (This cone Γ is a fundamental subject used in the classical harmonic analysis, see [6], [5], [20], [32] or any book on Hardy spaces). The presence of $\{1,2\}$ corresponds to our two variables x, y, see below. We then denote them by $L^p(\mathcal{M}, L_c^2(\widetilde{\Gamma}))$ (resp. $L^p(\mathcal{M}, L_r^2(\widetilde{\Gamma}))$). For simplicity, we will abbreviate them as $L^p(\mathcal{M}, L_c^2)$ (resp. $L^p(\mathcal{M}, L_r^2)$) if no confusion can arise.

2. Operator valued Hardy spaces

Let $1 \leq p < \infty$. For any $\mathcal{S}_\mathcal{M}$-valued simple function f on $\mathbb{R}$, we also use f to denote its Poisson integral on the upper half plane $\mathbb{R}_+^2 = \{(x,y) | y > 0\}$,

$$f(x,y) = \int_\mathbb{R} P_y(x-s)f(s)ds, \quad (x,y) \in \mathbb{R}_+^2,$$

where $P_y(x)$ is the Poisson kernel (i.e. $P_y(x) = \frac{1}{\pi}\frac{y}{x^2+y^2}$). Note that $f(x,y)$ is a harmonic function still with values in $\mathcal{S}_\mathcal{M}$, and so in $\mathcal{M}$. Define the $\mathcal{H}_c^p(\mathbb{R}, \mathcal{M})$ norm of f by

$$\|f\|_{\mathcal{H}_c^p} = \|\nabla f(x+t,y)\chi_\Gamma(x,y)\|_{L^p(L^\infty(\mathbb{R},dt)\otimes\mathcal{M}, L_c^2(\widetilde{\Gamma}))},$$

where ∇f is the gradient of the Poisson integral $f(x,y)$ and $\widetilde{\Gamma}$ is defined as in the end of Section 1.1. In this paper, we will always regard $\nabla f(x+t,y)\chi_\Gamma(x,y)$ as functions defined on $\mathbb{R} \times \widetilde{\Gamma}$ with $t \in \mathbb{R}, (x,y) \in \Gamma$ and

$$\nabla f(x+t,y)(1) = \frac{\partial f}{\partial x}(x+t,y), \quad \nabla f(x+t,y)(2) = \frac{\partial f}{\partial y}(x+t,y).$$

Set

$$|\nabla f(x+t,y)|^2 = |\frac{\partial f}{\partial x}(x+t,y)|^2 + |\frac{\partial f}{\partial y}(x+t,y)|^2.$$

Define the $\mathcal{H}_r^p(\mathbb{R}, \mathcal{M})$ norm of f by

$$\|f\|_{\mathcal{H}_r^p} = \|\nabla f(x+t,y)\chi_\Gamma\|_{L^p(L^\infty(\mathbb{R})\otimes\mathcal{M}, L_r^2)}.$$

Set $\mathcal{H}_c^p(\mathbb{R}, \mathcal{M})$ (resp. $\mathcal{H}_r^p(\mathbb{R}, \mathcal{M})$) to be the completion of the space of all $\mathcal{S}_\mathcal{M}$-valued simple function f's with finite $\mathcal{H}_c^p(\mathbb{R}, \mathcal{M})$(resp. $\mathcal{H}_r^p(\mathbb{R}, \mathcal{M})$) norm. Equipped respectively with the previous norms, $\mathcal{H}_c^p(\mathbb{R}, \mathcal{M})$ and $\mathcal{H}_r^p(\mathbb{R}, \mathcal{M})$ are Banach spaces. Define the noncommutative analogues of the classical Lusin integral by

$$(1.4) \qquad S_c(f)(t) = (\iint_\Gamma |\nabla f(x+t,y)|^2 dxdy)^{\frac{1}{2}}$$

$$(1.5) \qquad S_r(f)(t) = (\iint_\Gamma |\nabla f^*(x+t,y)|^2 dxdy)^{\frac{1}{2}}.$$

Note that

$$|\nabla f(x,y)|^2 = \int_{\{1,2\}} |\nabla f(x,y)(i)|^2 d\sigma(i).$$

Then, for $f \in \mathcal{H}_c^p(\mathbb{R},\mathcal{M})$,
$$\|f\|_{\mathcal{H}_c^p} = \|S_c(f)\|_{L^p(L^\infty(\mathbb{R})\otimes\mathcal{M})}$$
and the similar equality holds for $\mathcal{H}_r^p(\mathbb{R},\mathcal{M})$. $S_c(f)$ and $S_r(f)$ are the noncommutative analogues of the classical Lusin square function. We will need the noncommutative analogues of the classical Littlewood-Paley g-function, which are defined by

$$(1.6) \qquad G_c(f)(t) = (\int_{\mathbb{R}_+} |\nabla f(t,y)|^2 y dy)^{\frac{1}{2}}$$

$$(1.7) \qquad G_r(f)(t) = (\int_{\mathbb{R}_+} |\nabla f^*(t,y)|^2 y dy)^{\frac{1}{2}}$$

We will see, in Chapters 2 and 4, that
$$\|S_c(f)\|_{L^p(L^\infty(\mathbb{R})\otimes\mathcal{M})} \simeq \|G_c(f)\|_{L^p(L^\infty(\mathbb{R})\otimes\mathcal{M})}$$
$$\|S_r(f)\|_{L^p(L^\infty(\mathbb{R})\otimes\mathcal{M})} \simeq \|G_r(f)\|_{L^p(L^\infty(\mathbb{R})\otimes\mathcal{M})}$$
for all $1 \leq p < \infty$.

Define the Hardy spaces of noncommutative functions f as follows: if $1 \leq p < 2$,

$$(1.8) \qquad \mathcal{H}_{cr}^p(\mathbb{R},\mathcal{M}) = \mathcal{H}_c^p(\mathbb{R},\mathcal{M}) + \mathcal{H}_r^p(\mathbb{R},\mathcal{M})$$

equipped with the norm
$$\|f\|_{\mathcal{H}_{cr}^p} = \inf\{\|g\|_{\mathcal{H}_c^p} + \|h\|_{\mathcal{H}_r^p} : f = g+h, g \in \mathcal{H}_c^p(\mathbb{R},\mathcal{M}), h \in \mathcal{H}_r^p(\mathbb{R},\mathcal{M})\}$$
and if $2 \leq p < \infty$,

$$(1.9) \qquad \mathcal{H}_{cr}^p(\mathbb{R},\mathcal{M}) = \mathcal{H}_c^p(\mathbb{R},\mathcal{M}) \cap \mathcal{H}_r^p(\mathbb{R},\mathcal{M})$$

equipped with the norm
$$\|f\|_{\mathcal{H}_{cr}^p} = \max\{\|f\|_{\mathcal{H}_c^p}, \|f\|_{\mathcal{H}_r^p}\}.$$

Remark. We have
$$\mathcal{H}_c^2(\mathbb{R},\mathcal{M}) = \mathcal{H}_r^2(\mathbb{R},\mathcal{M}) = \mathcal{H}_{cr}^2(\mathbb{R},\mathcal{M}) = L^2(L^\infty(\mathbb{R})\otimes\mathcal{M}).$$
In fact, notice that $\triangle|f|^2 = 2|\nabla f|^2$ and $f(x,y)(|x|+y) \to 0$, $\nabla f(x,y)(|x|+y)^2 \to 0$ as $|x|+y \to 0$, for $S_\mathcal{M}$-valued simple function f's. By Green's theorem

$$\|\nabla f(t+x,y)\chi_\Gamma\|_{L^2(L^\infty(\mathbb{R})\otimes\mathcal{M},L_c^2)}^2$$
$$= 2\tau \iint_{\mathbb{R}_+^2} |\nabla f|^2 y dx dy$$
$$= \tau \iint_{\mathbb{R}_+^2} \triangle|f|^2 y dx dy$$
$$(1.10) \qquad = \tau \int_{\mathbb{R}} |f|^2 ds = \|f\|_{L^2(L^\infty(\mathbb{R})\otimes\mathcal{M})}^2.$$

Similarly, $\|f\|_{\mathcal{H}_r^2} = \|f^*\|_{L^2(L^\infty(\mathbb{R})\otimes\mathcal{M})} = \|f\|_{L^2(L^\infty(\mathbb{R})\otimes\mathcal{M})}$.

Note we have also the following polarized version of (1.10),

$$(1.11) \qquad 2\iint_{\mathbb{R}^2_+} \nabla f(x,y)\nabla g(x,y) y dx dy = \int_{\mathbb{R}} f(s)g(s)ds$$

for $S_{\mathcal{M}}$-valued simple function f,g's.

We will repeatedly use the following consequence of the convexity of the operator valued function: $x \mapsto |x|^2$ (This convexity follows from the convexity of $x \mapsto \langle x^*xh, h\rangle = \|xh\|^2$ for any h). Letting $f: (\Omega, \mu) \to \mathcal{M}$ be a weak-* integrable function, we have

$$(1.12) \qquad |\int_A f(t)d\mu(t)|^2 \leq \mu(A)\int_A |f(t)|^2 d\mu(t), \quad \forall A \subset \Omega$$

In particular, set $d\mu(t) = g^2(t)dt$,

$$(1.13) \qquad |\int_A f(t)g(t)dt|^2 \leq \int_A |f(t)|^2 dt \int_A g^2(t)dt, \quad \forall A \subset \mathbb{R}$$

for every measurable function g on $\mathbb{R}$, and

$$(1.14) \qquad |\int_A f(t)dt|^2 \leq \int_A |f(t)|^2 g^{-1}(t)dt \int_A g(t)dt, \quad \forall A \subset \mathbb{R}$$

for every positive measurable function g on $\mathbb{R}$.

Let $H^p(\mathbb{R})$ ($1 \leq p < \infty$) denote the classical Hardy space on $\mathbb{R}$. It is well known that

$$H^p(\mathbb{R}) = \{f \in L^p(\mathbb{R}) : S(f) \in L^p(\mathbb{R})\},$$

where $S(f)$ is the classical Lusin integral function ($S(f)$ is equal to $S_c(f)$ above by taking $\mathcal{M} = \mathbb{C}$). In the following, $H^p(\mathbb{R})$ is always equipped with the norm $\|S(f)\|_{L^p(\mathbb{R})}$.

PROPOSITION 1.2. *Let $1 \leq p < \infty$, $f \in \mathcal{H}_c^p(\mathbb{R}, \mathcal{M})$ and $m \in L^q(\mathcal{M})$ (with q the index conjugate to p). Then $\tau(mf) \in H^p(\mathbb{R})$ and*

$$\|\tau(mf)\|_{H^p} \leq \|m\|_{L^q(\mathcal{M})} \|f\|_{\mathcal{H}_c^p}.$$

Proof. Note that

$$\nabla(\tau(mf) * P) = \tau(m(f * \nabla P)) = \tau(m\nabla f),$$

here P is the Poisson kernel (i.e. $P_y(x) = \frac{1}{\pi}\frac{y}{x^2+y^2}$). By (1.13), we have

$$\|\tau(mf)\|_{H^p}^p$$
$$= \int_{\mathbb{R}} (\iint_{\Gamma} |\tau(m\nabla f(x+t,y))|^2 dxdy)^{\frac{p}{2}} dt$$

$$\leq \int_{\mathbb{R}} \sup_{\|g\|_{L^2(\tilde{\Gamma})}\leq 1} \left|\iint_{\Gamma} g\tau(m\nabla f(x+t,y)) dxdy\right|^p dt$$

$$= \int_{\mathbb{R}} \sup_{\|g\|_{L^2(\tilde{\Gamma})}\leq 1} \left|\tau\left[m \iint_{\Gamma} g_1 \frac{\partial f}{\partial x}(x+t,y) + g_2 \frac{\partial f}{\partial y}(x+t,y) dxdy\right]\right|^p dt$$

$$\leq \int_{\mathbb{R}} \sup_{\|g\|_{L^2(\tilde{\Gamma})}\leq 1} \|m\|_{L^q(\mathcal{M})}^p \left\|\iint_{\Gamma} g_1 \frac{\partial f}{\partial x}(x+t,y) + g_2 \frac{\partial f}{\partial y}(x+t,y) dxdy\right\|_{L^p(\mathcal{M})}^p dt$$

$$\leq \|m\|_{L^q(\mathcal{M})}^p \int_{\mathbb{R}} \sup_{\|g\|_{L^2(\tilde{\Gamma})}\leq 1} \left\|(\iint_{\Gamma}|g|^2 dxdy)^{\frac{1}{2}}(\iint_{\Gamma}|\nabla f(x+t,y)|^2 dxdy)^{\frac{1}{2}}\right\|_{L^p(\mathcal{M})}^p dt$$

$$\leq \|m\|_{L^q(\mathcal{M})}^p \tau \int_{\mathbb{R}} (\iint_{\Gamma}|\nabla f(x+t,y)|^2 dxdy)^{\frac{p}{2}} dt$$

$$= \|m\|_{L^q(\mathcal{M})}^p \|f\|_{\mathcal{H}_c^p}^p. \blacksquare$$

Remark. We should emphasize that for two functions g, f defined on $\tilde{\Gamma}$, we always set
$$gf(z) = g(z)(1)f(z)(1) + g(z)(2)f(z)(2).$$
Then in the above formula $|\tau(m\nabla f(x+t,y))|^2$ and $g\tau(m\nabla f(x+t,y))$ etc. are functions defined on Γ. We will use very often such a product for ($\mathcal{M}$-valued) functions defined on $\tilde{\Gamma}$.

Remark. (i) $\int f dt = 0, \forall f \in \mathcal{H}_c^1(\mathbb{R}, \mathcal{M})$. In fact, if $f \in \mathcal{H}_c^1(\mathbb{R}, \mathcal{M})$, by Proposition 1.2 and the classical property of H^1 (see [**32**], p.128), we have $\tau(m \int f dt) = 0, \forall m \in \mathcal{M}$. Thus $\int f dt = 0$.

(ii) The collection of all $S_\mathcal{M}$-valued simple functions f such that $\int f dt = 0$ is a dense subset of $\mathcal{H}_c^p(\mathbb{R}, \mathcal{M})(1 < p < \infty)$. Note that

$$\lim_{N\to\infty} \left\|\frac{m}{N}\chi_{[-N,N]}(t)\right\|_{\mathcal{H}_c^p(\mathbb{R},\mathcal{M})} = 0, \quad \forall m \in S_\mathcal{M}.$$

For a simple function f, let $f_N = f - \frac{\int f dt}{N}\chi_{[-N,N]}$. Then $\int f_N = 0$ and $f_N \to f$ in $\mathcal{H}_c^p(\mathbb{R}, \mathcal{M})$.

Remark. See [**5**] and [**32**] for discussions on the classical Lusin integral and the Littlewood-Paley g-function and the fact that a scalar valued function is in H^1 if and only if its Lusin integral is in L^1. We define the noncommutative Hardy spaces $\mathcal{H}_{cr}^p(\mathbb{R}, \mathcal{M})$ differently for the case $1 \leq p < 2$ and $p \geq 2$ (respectively by (1.8) and (1.9)) as Pisier and Xu did for noncommutative martingales in [**16**]. This is to get the expected equivalence between $\mathcal{H}_{cr}^p(\mathbb{R}, \mathcal{M})$ and $L^p(\mathbb{R}, \mathcal{M})$ for $1 < p < \infty$

(see Chapter 5). And $\mathcal{H}_c^p(\mathbb{R},\mathcal{M})$ or $\mathcal{H}_r^p(\mathbb{R},\mathcal{M})$ alone could be very far away from $L^p(\mathbb{R},\mathcal{M})$ for $p \neq 2$.

3. Operator valued BMO spaces

Now, we introduce the noncommutative analogue of BMO spaces. For any interval I on $\mathbb{R}$, we will denote its center by C_I and its Lebesgue measure by $|I|$. Let $\varphi \in L^\infty(\mathcal{M}, L_c^2(\mathbb{R}, \frac{dt}{1+t^2}))$. By Proposition 1.1 (and our convention), for every $g \in L^2(\mathbb{R}, \frac{dt}{1+t^2})$, $\int_\mathbb{R} g\varphi \frac{dt}{1+t^2} \in \mathcal{M}$. Then the mean value of φ over I $\varphi_I := \frac{1}{|I|}\int_I \varphi(s)ds$ exists as an element in $\mathcal{M}$. And the Poisson integral of φ

$$\varphi(x,y) = \int_\mathbb{R} P_y(x-s)\varphi(s)ds$$

also exists as an element in $\mathcal{M}$. Set

(1.15) $$\|\varphi\|_{\mathrm{BMO}_c} = \sup_{I \subset \mathbb{R}} \left\{ \left\|\left(\frac{1}{|I|}\int_I |\varphi - \varphi_I|^2 d\mu\right)^{\frac{1}{2}}\right\|_\mathcal{M} \right\}$$

where again $|\varphi-\varphi_I|^2 = (\varphi-\varphi_I)^*(\varphi-\varphi_I)$ and the supremum runs over all intervals $I \subset \mathbb{R}$.(see Let H be the Hilbert space on which $\mathcal{M}$ acts. Obviously, we have

(1.16) $$\|\varphi\|_{\mathrm{BMO}_c} = \sup_{e \in H, \|e\|=1} \|\varphi e\|_{\mathrm{BMO}_2(\mathbb{R},H)}$$

where $\mathrm{BMO}_2(\mathbb{R},H)$ is the usual H-valued BMO space on $\mathbb{R}$. Thus $\|\cdot\|_{\mathrm{BMO}_c}$ is a norm modulo constant functions. Set $\mathrm{BMO}_c(\mathbb{R},\mathcal{M})$ to be the space of all $\varphi \in L^\infty(\mathcal{M}, L_c^2(\mathbb{R}, \frac{dt}{1+t^2}))$ such that $\|\varphi\|_{\mathrm{BMO}_c} < \infty$. $\mathrm{BMO}_r(\mathbb{R},\mathcal{M})$ is defined as the space of all φ's such that $\varphi^* \in \mathrm{BMO}_c(\mathbb{R},\mathcal{M})$ with the norm $\|\varphi\|_{\mathrm{BMO}_r} = \|\varphi^*\|_{\mathrm{BMO}_c}$. We define $\mathrm{BMO}_{cr}(\mathbb{R},\mathcal{M})$ as the intersection of these two spaces

$$\mathrm{BMO}_{cr}(\mathbb{R},\mathcal{M}) = \mathrm{BMO}_c(\mathbb{R},\mathcal{M}) \cap \mathrm{BMO}_r(\mathbb{R},\mathcal{M})$$

with the norm

$$\|\varphi\|_{\mathrm{BMO}_{cr}} = \max\{\|\varphi\|_{\mathrm{BMO}_c}, \|\varphi\|_{\mathrm{BMO}_r}\}.$$

As usual, the constant functions are considered as zero in these BMO spaces, and then these spaces are normed spaces (modulo constants).

Given an interval I, we denote by $2^k I$ the interval $\{t : |t-C_I| < 2^{k-1}|I|\}$. The technique used in the proof of the following Proposition is classical (see [**32**]).

PROPOSITION 1.3. *Let $\varphi \in \mathrm{BMO}_c(\mathbb{R},\mathcal{M})$. Then*

$$\|\varphi\|_{L^\infty(\mathcal{M},L_c^2(\mathbb{R},\frac{dt}{1+t^2}))} \leq c(\|\varphi\|_{\mathrm{BMO}c} + \|\varphi_{I_1}\|_\mathcal{M})$$

where $I_1 = (-1,1]$. Moreover, $\mathrm{BMO}_c(\mathbb{R},\mathcal{M}), \mathrm{BMO}_r(\mathbb{R},\mathcal{M}), \mathrm{BMO}_{cr}(\mathbb{R},\mathcal{M})$ are Banach spaces.

Proof. Let $\varphi \in \mathrm{BMO}_c(\mathbb{R}, \mathcal{M})$ and I be an interval. Using (1.12), (1.14) we have

$$|\varphi_{2^n I} - \varphi_I|^2 \leq n \sum_{k=0}^{n-1} |\varphi_{2^k I} - \varphi_{2^{k+1} I}|^2$$

$$= n \sum_{k=0}^{n-1} \left| \frac{1}{|2^k I|} \int_{2^k I} (\varphi(s) - \varphi_{2^{k+1} I}) ds \right|^2$$

$$\leq n \sum_{k=0}^{n-1} \frac{2}{|2^{k+1} I|} \int_{2^{k+1} I} |\varphi(s) - \varphi_{2^{k+1} I}|^2 ds$$

(1.17) $$\leq 2n \|\varphi\|_{\mathrm{BMO}_c}^2.$$

By (1.14), (1.17),

$$\left\| \int_{\mathbb{R}} \frac{|\varphi(t)|^2}{1+t^2} dt \right\|_{\mathcal{M}}$$

$$= \left\| \int_{I_1} \frac{|\varphi(t)|^2}{1+t^2} dt + \sum_{k=0}^{\infty} \int_{2^{k+1} I_1 / 2^k I_1} \frac{|\varphi(t)|^2}{1+t^2} dt \right\|_{\mathcal{M}}$$

$$\leq 2 \left\| \int_{I_1} (|\varphi(t) - \varphi_{I_1}|^2 + |\varphi_{I_1}|^2) dt \right\|_{\mathcal{M}}$$

$$+ 4 \left\| \sum_{k=0}^{\infty} \int_{2^{k+1} I_1 / 2^k I_1} \frac{|\varphi(t) - \varphi_{2^{k+1} I_1}|^2 + |\varphi_{2^{k+1} I_1} - \varphi_{I_1}|^2 + |\varphi_{I_1}|^2}{2^{2k}} dt \right\|_{\mathcal{M}}$$

(1.18) $$\leq c(\||\varphi_{I_1}|^2\|_{\mathcal{M}} + \|\varphi\|_{\mathrm{BMO}_c}^2)$$

Thus

$$\|\varphi\|_{L^\infty(\mathcal{M}, L_c^2(\mathbb{R}, \frac{dt}{1+t^2}))} = \left\| \left(\int_{\mathbb{R}} \frac{|\varphi(t)|^2}{1+t^2} dt \right)^{\frac{1}{2}} \right\|_{\mathcal{M}} \leq c(\|\varphi_{I_1}\|_{\mathcal{M}} + \|\varphi\|_{\mathrm{BMO}_c})$$

And then $\mathrm{BMO}_c(\mathbb{R}, \mathcal{M})$ is complete. Consequently, $\mathrm{BMO}_c(\mathbb{R}, \mathcal{M})$, $\mathrm{BMO}_r(\mathbb{R}, \mathcal{M})$, $\mathrm{BMO}_{cr}(\mathbb{R}, \mathcal{M})$ are Banach spaces. ∎

It is classical that BMO functions are related with Carleson measures(See [6], [20]). The same relation still holds in the present noncommutative setting. We say that an $\mathcal{M}$-valued measure $d\lambda$ on $\mathbb{R}_+^2$ is a Carleson measure if

$$N(\lambda) = \sup_I \left\{ \frac{1}{|I|} \left\| \iint_{T(I)} d\lambda \right\|_{\mathcal{M}} : I \in \mathbb{R} \text{ interval} \right\} < \infty,$$

where, as usual, $T(I) = I \times (0, |I|]$.

LEMMA 1.4. *Let $\varphi \in \mathrm{BMO}_c(\mathbb{R}, \mathcal{M})$. Then $d\lambda_\varphi = |\nabla \varphi|^2 y dx dy$ is an $\mathcal{M}$-valued Carleson measure on $\mathbb{R}_+^2$ and $N(\lambda_\varphi) \leq c \|\varphi\|_{\mathrm{BMO}_c}^2$.*

Proof. The proof is very similar to the scalar situation (see [32], p.160). For any interval I on $\mathbb{R}$, write $\varphi = \varphi_1 + \varphi_2 + \varphi_3$, where $\varphi_1 = (\varphi - \varphi_{2I})\chi_{2I}, \varphi_2 = (\varphi - \varphi_{2I})\chi_{(2I)^c}$ and $\varphi_3 = \varphi_{2I}$. Set

$$d\lambda_{\varphi_1} = |\nabla \varphi_1|^2 y dx dy, d\lambda_{\varphi_2} = |\nabla \varphi_2|^2 y dx dy.$$

Thus
$$N(\lambda_\varphi) \leq 2(N(\lambda_{\varphi_1}) + N(\lambda_{\varphi_2})).$$
We treat $N(\lambda_{\varphi_1})$ first. Notice that $\triangle|\varphi_1|^2 = 2|\nabla\varphi_1|^2$ and $\varphi_1(x,y)(|x|+y) \to 0, \nabla\varphi_1(x,y)(|x|+y)^2 \to 0$ as $|x|+y \to 0$. By Green's theorem

$$(1.19) \quad \frac{1}{|I|}\left\|\iint_{T(I)} |\nabla\varphi_1|^2 y\, dx dy\right\|_{\mathcal{M}} \leq \frac{1}{|I|}\left\|\iint_{\mathbb{R}_2^+} |\nabla\varphi_1|^2 y\, dx dy\right\|_{\mathcal{M}}$$
$$= \frac{1}{2|I|}\left\|\int_{\mathbb{R}} |\varphi_1|^2 ds\right\|_{\mathcal{M}}$$
$$= \frac{1}{2|I|}\left\|\int_{2I} |\varphi - \varphi_{2I}|^2 ds\right\|_{\mathcal{M}} \leq \|\varphi\|_{\mathrm{BMO}_c}^2$$

To estimate $N(\lambda_{\varphi_1})$, we note
$$|\nabla P_y(x-s)|^2 \leq \frac{1}{4(x-s)^4} \leq \frac{1}{4|I|^4 2^{4k}}, \quad \forall s \in 2^{k+1}I/2^kI, \quad (x,y) \in T(I),$$
by (1.14) and (1.17)

$$\frac{1}{|I|}\left\|\iint_{T(I)} |\nabla\varphi_2|^2 y\, dx dy\right\|_{\mathcal{M}}$$
$$= \frac{1}{|I|}\left\|\iint_{T(I)} |\nabla \int_{-\infty}^{+\infty} P_y(x-s)\varphi_2(s)ds|^2 y\, dx dy\right\|_{\mathcal{M}}$$
$$\leq \frac{1}{|I|}\iint_{T(I)} \sum_{k=1}^\infty \int_{2^{k+1}I/2^kI} |\nabla P_y(x-s)|^2 2^{2k} ds \sum_{k=1}^\infty \frac{1}{2^{2k}}\left\|\int_{2^{k+1}I} |\varphi_2|^2 ds\right\|_{\mathcal{M}} y\, dx dy$$
$$\leq \frac{c}{|I|}\iint_{T(I)} \frac{1}{|I|^2}\|\varphi\|_{\mathrm{BMO}_c}^2 y\, dx dy \leq c\|\varphi\|_{\mathrm{BMO}_c}^2$$

Therefore $N(\lambda_{\varphi_i}) \leq c\|\varphi\|_{\mathrm{BMO}_c}^2, i = 1, 2$, and then $N(\lambda_\varphi) \leq c\|\varphi\|_{\mathrm{BMO}_c}^2$. ∎

Remark. We will see later (Corollary 2.6) that the converse to lemma 1.4 is also true.

We will need the following elementary fact to make our later applications of Green's theorem rigorous in Chapters 2 and 4.

LEMMA 1.5. *Suppose $\varphi \in \mathrm{BMO}_c(\mathbb{R}, \mathcal{M})$ and suppose I is an interval such that $\varphi_I = 0$. Let $3I$ be the interval concentric with I having length $3|I|$. Then there is $\psi \in \mathrm{BMO}_c(\mathbb{R}, \mathcal{M})$ such that $\psi = \varphi$ on $I, \psi = 0$ on $\mathbb{R}\setminus 3I$ and*
$$\|\psi\|_{\mathrm{BMO}_c} \leq c\|\varphi\|_{\mathrm{BMO}_c}.$$

Proof. This is well known for the classical BMO and a proof is outlined in [6], p. 269. One can check that the method to construct ψ mentioned there works as well for $\mathrm{BMO}_c(\mathbb{R}, \mathcal{M})$. ∎

Remark. We have seen that the noncommutative $\mathrm{BMO}_c(\mathbb{R}, \mathcal{M})$ are well adapted to many generalizations of classical results, such as Proposition 1.3 and Lemma

1.4, 1.5. We will also prove an analogue of the classical Fefferman duality between $\mathcal{H}^1$ and BMO in the next chapter. However, unlike the classical case, we could not replace the power 2 by p in the definition of the noncommutative BMO norm ((1.15)). In fact, $\sup_{I\subset\mathbb{R}} \left\| \left(\frac{1}{|I|} \int_I |\varphi - \varphi_I|^p d\mu \right)^{\frac{1}{p}} \right\|_{\mathcal{M}}$ may not be a norm for $p \neq 2$ in the noncommutative case (Note we do not have $|x_1 + x_2| \leq |x_1| + |x_2|$ in general for $x_1, x_2 \in \mathcal{M}$). See the remark at the end of Chapter 6 for more information.

CHAPTER 2

The Duality between $\mathcal{H}^1$ and BMO

The main result (Theorem 2.4) of this chapter is the analogue in our setting of the famous Fefferman duality theorem between H^1 and BMO.

1. The bounded map from $L^\infty(L^\infty(\mathbb{R}) \otimes \mathcal{M}, L_c^2)$ to $\mathrm{BMO}_c(\mathbb{R}, \mathcal{M})$

As in the classical case, we will embed $\mathcal{H}_c^1(\mathbb{R}, \mathcal{M})$ into a larger space $L^1(L^\infty(\mathbb{R}) \otimes \mathcal{M}, L_c^2)$, which requires the following maps Φ, Ψ.

DEFINITION. *We define a map Φ from $\mathcal{H}_c^p(\mathbb{R}, \mathcal{M})$ $(1 \leq p < \infty)$ to $L^p(L^\infty(\mathbb{R}) \otimes \mathcal{M}, L_c^2(\widetilde{\Gamma}))$ by*
$$\Phi(f)(x, y, t) = \nabla f(x+t, y) \chi_\Gamma(x, y)$$
and a map Ψ for a sufficiently nice $h \in L^p(L^\infty(\mathbb{R}) \otimes \mathcal{M}, L_c^2(\widetilde{\Gamma}))$ $(1 \leq p \leq \infty)$ by

$$(2.1) \quad \Psi(h)(s) = \int_\mathbb{R} \iint_\Gamma h(x, y, t) Q_y(x+t-s) dy dx dt; \quad \forall s \in \mathbb{R}$$

where, $Q_y(x)$ is defined as a function on $\mathbb{R} \times \widetilde{\Gamma}$ by

$$(2.2) \quad Q_y(x)(1) = \frac{\partial P_y(x)}{\partial x}, \quad Q_y(x)(2) = \frac{\partial P_y(x)}{\partial y}; \forall (x, y) \in \Gamma.$$

Note that Φ is simply the natural embedding of $\mathcal{H}_c^p(\mathbb{R}, \mathcal{M})$ into $L^p(L^\infty(\mathbb{R}) \otimes \mathcal{M}, L_c^2(\widetilde{\Gamma}))$. On the other hand, Ψ is well defined for sufficiently nice h, more precisely "nice" will mean that $h(x, y, t) = \sum_{i=1}^n m_i f_i(t) \chi_{A_i}$ with $m_i \in S_\mathcal{M}, A_i \in \widetilde{\Gamma}, |A_i| < \infty$ and with scalar valued simple functions f_i. In this case, it is easy to check that $\Psi(h) \in L^p(\mathcal{M}, L_c^2(\mathbb{R}, \frac{dt}{1+t^2}))$.

We will prove that Ψ extends to a bounded map from $L^\infty(L^\infty(\mathbb{R}) \otimes \mathcal{M}, L_c^2(\widetilde{\Gamma}))$ to $\mathrm{BMO}_c(\mathbb{R}, \mathcal{M})$ (see Lemma 2.2) and also from $L^p(L^\infty(\mathbb{R}) \otimes \mathcal{M}, L_c^2(\widetilde{\Gamma}))$ to $\mathcal{H}_c^p(\mathbb{R}, \mathcal{M})$ for all $1 < p < \infty$ (see Theorem 4.8). The following proposition, combined with Theorem 4.8 in Chapter 4, implies that Ψ is a projection of $L^p(L^\infty(\mathbb{R}) \otimes \mathcal{M}, L_c^2(\widetilde{\Gamma}))$ onto $\mathcal{H}_c^p(\mathbb{R}, \mathcal{M})$ if we identify $\mathcal{H}_c^p(\mathbb{R}, \mathcal{M})$ with a subspace of $L^p(L^\infty(\mathbb{R}) \otimes \mathcal{M}, L_c^2(\widetilde{\Gamma}))$ via Φ.

PROPOSITION 2.1. *For any $f \in \mathcal{H}_c^p(\mathbb{R}, \mathcal{M})$ $(1 \leq p < \infty)$,*
$$\Psi \Phi(f) = f$$

Proof. We have
$$\int_{-\infty}^{+\infty} \iint_\Gamma \Phi(f) \nabla g(t+x, y) dy dx dt$$
$$= \int_{-\infty}^{+\infty} \int_{-\infty}^{+\infty} \iint_\Gamma \Phi(f) Q_y(x+t-s) dy dx dt g(s) ds.$$

On the other hand, by (1.11) we have

$$\int_{-\infty}^{+\infty} \iint_{\Gamma} \Phi(f) \nabla g(t+x,y) dy dx dt = \int_{-\infty}^{+\infty} f(s) g(s) ds$$

for every g good enough. Therefore

$$\int_{-\infty}^{+\infty} \iint_{\Gamma} \Phi(f) Q_y(x+t-s) dy dx dt = f(s)$$

almost everywhere. This is $\Psi\Phi(f) = f$. ∎

We can also prove $\Psi\Phi(\varphi) = \varphi$ by showing directly the Poisson integral of $\Psi\Phi(\varphi)$ coincides with that of φ, namely

$$(2.3) \quad \int_{\mathbb{R}} \Psi\Phi(\varphi)(w) P_v(u-w) dw = \int_{\mathbb{R}} \varphi(w) P_v(u-w) dw, \quad \forall (u,v) \in \mathbb{R}_+^2.$$

Indeed, using elementary properties of the Poisson kernel, we have

$$\int_{\mathbb{R}} \Psi\Phi(\varphi)(h) P_v(u-h) dh$$
$$= \int_{\mathbb{R}} \int_{\mathbb{R}} \iint_{\Gamma} \int_{\mathbb{R}} \varphi(s) \nabla P_y(x+t-s) ds \nabla P_y(x+t-h) dy dx dt P_v(u-h) dh$$
$$= \int_{\mathbb{R}} \varphi(s) \iint_{\Gamma} \int_{\mathbb{R}} \int_{\mathbb{R}} \frac{\partial}{\partial y} P_y(x+t-s) \frac{\partial}{\partial y} P_y(x+t-h) P_v(u-h) dt dh dx dy ds$$
$$+ \int_{\mathbb{R}} \varphi(s) \iint_{\Gamma} \int_{\mathbb{R}} \int_{\mathbb{R}} \frac{\partial}{\partial x} P_y(x+t-s) \frac{\partial}{\partial x} P_y(x+t-h) P_v(u-h) dt dh dx dy ds$$
$$= \int_{\mathbb{R}} \varphi(s) \int_{\mathbb{R}} \iint_{\mathbb{R}_+^2} \frac{\partial}{\partial y} P_y(x-s) \frac{\partial}{\partial y} P_y(x-h) 2y dy dx P_v(u-h) dh ds$$
$$+ \int_{\mathbb{R}} \varphi(s) \int_{\mathbb{R}} \int_{\mathbb{R}} \frac{\partial}{\partial s} P_y(x-s) \frac{\partial}{\partial u} P_{y+v}(x-u) 2y dx dy ds$$
$$= \int_{\mathbb{R}} \varphi(s) \int_0^{\infty} 2y \frac{\partial^2}{\partial v^2} P_{v+2y}(u-s) dy ds - \int_{\mathbb{R}} \varphi(s) \int_0^{\infty} 2y \frac{\partial^2}{\partial u^2} P_{v+2y}(u-s) dy ds$$
$$= \int_{\mathbb{R}} \varphi(s) \int_0^{\infty} y \frac{\partial^2}{\partial y^2} P_{v+2y}(u-s) dy ds$$
$$= \int_{\mathbb{R}} \varphi(s) (0 - \int_0^{\infty} \frac{\partial}{\partial y} P_{v+2y}(u-s) dy) ds$$
$$= \int_{\mathbb{R}} \varphi(s) P_v(u-s) ds. \quad \blacksquare$$

LEMMA 2.2. Ψ *extends to a bounded map from* $L^{\infty}(L^{\infty}(\mathbb{R}) \otimes \mathcal{M}, L_c^2(\widetilde{\Gamma}))$ *to* $\mathrm{BMO}_c(\mathbb{R}, \mathcal{M})$ *of norm controlled by a universal constant.*

Proof. Let $\mathcal{S}$ be the family of all $L^{\infty}(\mathbb{R}) \otimes \mathcal{M}$-valued simple functions h which can written as $h(x,y,t) = \sum_{i=1}^n m_i f_i(t) \chi_{A_i}(x,y)$ with $m_i \in S_{\mathcal{M}}$, $f_i \in L^{\infty}(\mathbb{R}) \cap L^1(\mathbb{R})$ and compact $A_i \subset \widetilde{\Gamma}$. (By compact A_i we mean that the two components of A_i are compact subsets in Γ.) Note that $\mathcal{S}$ is w*-dense in $L^{\infty}(L^{\infty}(\mathbb{R}) \otimes \mathcal{M}, L_c^2(\widetilde{\Gamma}))$ (in fact, the unit ball of $\mathcal{S}$ is w*-dense in the unit ball of $L^\infty(L^\infty(\mathbb{R})\overline{\otimes}\mathcal{M}, L_c^2(\widetilde{\Gamma})))$. We will first show that

$$\|\Psi(h)\|_{\mathrm{BMO}_c} \leq c\|h\|_{L^\infty(L^\infty(\mathbb{R})\overline{\otimes}\mathcal{M}, L_c^2)} \ , \ \forall\, h \in \mathcal{S}. \tag{2.4}$$

Fix $h \in \mathcal{S}$ and let $\varphi = \Psi(h)$. Then $\varphi \in L^\infty(\mathcal{M}, L_c^2(\mathbb{R}, \frac{dt}{1+t^2}))$ by Proposition 1.3. To estimate the BMO_c-norm of φ, we fix an interval I and set $h = h_1 + h_2$ with

$$\begin{aligned} h_1(x,y,t) &= h(x,y,t)\chi_{2I}(t) \\ h_2(x,y,t) &= h(x,y,t)\chi_{(2I)^c}(t). \end{aligned}$$

Let

$$B_I = \int_{-\infty}^{+\infty}\iint_\Gamma Q_I h_2\, dy dx dt$$

with the notation $Q_I(x,t) = \frac{1}{|I|}\int_I Q_y(x+t-s)ds$. Now

$$\begin{aligned} &\frac{1}{|I|}\int_I |\varphi(s) - B_I|^2 ds \\ &\leq \frac{2}{|I|}\int_I |\int_{(2I)^c}\iint_\Gamma (Q_y(x+t-s) - Q_I)h\, dx dy dt|^2 ds \\ &\quad + \frac{2}{|I|}\int_I |\int_{-\infty}^{+\infty}\iint_\Gamma Q_y(x+t-s)h_1\, dx dy dt|^2 ds \\ &= A + B \end{aligned}$$

Notice that

$$\iint_\Gamma |Q_y(x+t-s) - Q_I|^2 dxdy \leq c\iint_\Gamma (\frac{|I|}{(|x+t-s|+y)^3})^2 dxdy$$
$$\leq c|I|^2(t-C_I)^{-4} \tag{2.5}$$

for every $t \in (2I)^c$ and $s \in I$. By (1.14)

$$\left|\iint_\Gamma (Q_y(x+t-s) - Q_I)h\, dxdy\right|^2 \leq c|I|^2(t-C_I)^{-4}\iint_\Gamma h^*h\, dxdy$$

and by (1.14) again,

$$\begin{aligned} &\|A\|_\mathcal{M} \\ &\leq c\|\int_{(2I)^c}(t-C_I)^{-2}dt \int_{(2I)^c}(t-C_I)^2\iint_\Gamma h^*h\, dxdy|I|^2(t-C_I)^{-4}dt\|_\mathcal{M} \\ &\leq \|\frac{c}{|I|}\int_{(2I)^c}|I|^2(t-C_I)^{-2}\iint_\Gamma h^*h\, dxdydt\|_\mathcal{M} \\ &\leq c\|h\|^2_{L^\infty(L^\infty(\mathbb{R})\overline{\otimes}\mathcal{M}, L_c^2)} \end{aligned}$$

For the second term B, we have

$$\|B\|_{\mathcal{M}}$$
$$\leq \frac{2}{|I|}\|\int_{\mathbb{R}}|\int_{\mathbb{R}}\iint_{\Gamma}Q_y(x+t-s)h_1 dxdydt|^2 ds\|_{\mathcal{M}}$$
$$= \frac{2}{|I|}\sup_{\tau|a|=1}\tau(|a|\int_{\mathbb{R}}|\int_{\mathbb{R}}\iint_{\Gamma}Q_y(x+t-s)h_1 dxdydt|^2 ds)$$
$$= \frac{2}{|I|}\sup_{\tau|a|=1}\tau\int_{\mathbb{R}}|\int_{\mathbb{R}}\iint_{\Gamma}Q_y(x+t-s)h_1|a|^{\frac{1}{2}}dxdydt|^2 ds$$
$$= \frac{2}{|I|}\sup_{\tau|a|=1}\sup_{\|f\|_{L^2(L^\infty(\mathbb{R})\otimes\mathcal{M})}=1}(\tau\int_{\mathbb{R}}f(s)\int_{\mathbb{R}}\iint_{\Gamma}Q_y(x+t-s)h_1|a|^{\frac{1}{2}}dxdydtds)^2$$
$$= \frac{2}{|I|}\sup_{\tau|a|=1}\sup_{\|f\|_{L^2(L^\infty(\mathbb{R})\otimes\mathcal{M})}=1}(\tau\int_{\mathbb{R}}\iint_{\Gamma}\nabla f(t+x,y)h_1|a|^{\frac{1}{2}}dxdydt)^2$$

Hence by the Cauchy-Schwarz inequality and (1.10)

$$\|B\|_{\mathcal{M}} \leq \frac{2}{|I|}\sup_{\tau|a|=1}\tau\int_{\mathbb{R}}\iint_{\Gamma}h_1^*h_1|a|dxdydt$$
$$\leq \frac{2}{|I|}\|\int_{\mathbb{R}}\iint_{\Gamma}h_1^*h_1 dxdydt\|_{\mathcal{M}}$$
$$= \frac{2}{|I|}\|\int_{2I}\iint_{\Gamma}h^*h\,dxdydt\|_{\mathcal{M}}$$
$$\leq 4\|h\|^2_{L^\infty(L^\infty(\mathbb{R})\otimes\mathcal{M},L_c^2)}$$

Thus

$$\|\varphi\|_{\mathrm{BMO}_c} \leq c\|h\|_{L^\infty(L^\infty(\mathbb{R})\otimes\mathcal{M},L_c^2)}.$$

In particular, by Proposition 1.3,

$$\|\varphi\|_{L^\infty(\mathcal{M},L_c^2(\mathbb{R},\frac{dt}{1+t^2}))} \leq c\|h\|_{L^\infty(L^\infty(\mathbb{R})\otimes\mathcal{M},L_c^2)}.$$

Thus we have proved the boundedness of Ψ from the w*-dense vector subspace $\mathcal{S}$ of $L^\infty(L^\infty(\mathbb{R})\otimes\mathcal{M},L_c^2(\widetilde{\Gamma}))$ to $\mathrm{BMO}_c(\mathbb{R},\mathcal{M})$. Now we extend Ψ to the whole $L^\infty(L^\infty(\mathbb{R})\otimes\mathcal{M},L_c^2(\widetilde{\Gamma}))$. To this end we first extend Ψ to a bounded map from $L^\infty(L^\infty(\mathbb{R})\otimes\mathcal{M},L_c^2(\widetilde{\Gamma}))$ into $L^\infty(\mathcal{M},L_c^2(\mathbb{R},\frac{dt}{1+t^2}))$. By the discussion above, Ψ is also bounded from $\mathcal{S}$ to $L^\infty(\mathcal{M},L_c^2(\mathbb{R},\frac{dt}{1+t^2}))$. Let H_0^1 be the subspace of all $f \in H^1(\mathbb{R})$ such that $(1+t^2)f(t) \in L^2(\mathbb{R})$. Let $L^1(\mathcal{M}) \otimes H_0^1$ denote the algebraic tensor product of $L^1(\mathcal{M})$ and H_0^1. Note that

$$L^1(\mathcal{M}) \otimes H_0^1 \subset \mathcal{H}_c^1(\mathbb{R},\mathcal{M}), \quad L^1(\mathcal{M}) \otimes H_0^1 \subset L^1(\mathcal{M},L_c^2(\mathbb{R},\frac{dt}{1+t^2}))$$

and $L^1(\mathcal{M}) \otimes H_0^1$ is dense in both of the latter spaces. Moreover, it is easy to see that for any $h \in \mathcal{S}$ and $f \in L^1(\mathcal{M}) \otimes H_0^1$

$$\tau\int_{-\infty}^{+\infty}\iint_{\Gamma}h^*(x,y,t)\nabla f(t+x,y)dydxdt = \tau\int_{-\infty}^{+\infty}\Psi(h)^*(s)f(s)ds.$$

Then it follows that Ψ is continuous from $(\mathcal{S}, \sigma(\mathcal{S}, L^1(L^\infty(\mathbb{R}) \otimes \mathcal{M}, L_c^2(\widetilde{\Gamma}))))$ to $(L^\infty(\mathcal{M}, L_c^2(\mathbb{R}, \frac{dt}{1+t^2})), \sigma(L^\infty(\mathcal{M}, L_c^2(\mathbb{R}, \frac{dt}{1+t^2})), L^1(\mathcal{M}) \otimes H_0^1))$.

Now given $f \in L^1(\mathcal{M}) \otimes H_0^1$ we define $\Psi_*(f) : \mathcal{S} \to \mathbb{C}$ by

$$\Psi_*(f)(h) = \tau \int_{-\infty}^{+\infty} \Psi(h)^*(s) f(s) ds.$$

Then $\Psi_*(f)$ is an anti-linear functional on $\mathcal{S}$ and is continuous with respect to the w*-topology; hence $\Psi_*(f)$ extends to a w*-continuous anti-linear functional on $L^\infty(L^\infty(\mathbb{R}) \otimes \mathcal{M}, L_c^2(\widetilde{\Gamma})))$, i.e. an element in $L^1(L^\infty(\mathbb{R}) \otimes \mathcal{M}, L_c^2(\widetilde{\Gamma}))$, still denoted by $\Psi_*(f)$. By the w*-density of $\mathcal{S}$ in $L^\infty(L^\infty(\mathbb{R}) \otimes \mathcal{M}, L_c^2(\widetilde{\Gamma})))$, this extension is unique. Therefore, we have defined a map

$$\Psi_* : L^1(\mathcal{M}) \otimes H_0^1 \to L^1(L^\infty(\mathbb{R}) \otimes \mathcal{M}, L_c^2(\widetilde{\Gamma})).$$

The above uniqueness of the extension $\Psi_*(f)$ for any given f implies that Ψ_* is linear. On the other hand, by what we already proved in the previous part, we have

$$\begin{aligned}
|\Psi_*(f)(h)| &\leq \|f\|_{L^1(\mathcal{M}, L_c^2(\mathbb{R}, \frac{dt}{1+t^2}))} \|\Psi(h)\|_{L^\infty(\mathcal{M}, L_c^2(\mathbb{R}, \frac{dt}{1+t^2}))} \\
&\leq c \|f\|_{L^1(\mathcal{M}, L_c^2(\mathbb{R}, \frac{dt}{1+t^2}))} \|h\|_{L^\infty(L^\infty(\mathbb{R}) \otimes \mathcal{M}, L_c^2)}.
\end{aligned}$$

Since the unit ball of $\mathcal{S}$ is w*-dense in the unit ball of $L^\infty(L^\infty(\mathbb{R}) \otimes \mathcal{M}, L_c^2(\widetilde{\Gamma})))$, it follows that

$$\Psi_* : (L^1(\mathcal{M}) \otimes H_0^1, \|\cdot\|_{L^1(\mathcal{M}, L_c^2(\mathbb{R}, \frac{dt}{1+t^2})))}) \to L^1(L^\infty(\mathbb{R}) \otimes \mathcal{M}, L_c^2(\widetilde{\Gamma}))$$

is bounded and its norm is majorized by c. This, together with the density of $L^1(\mathcal{M}) \otimes H_0^1$ in $L^1(\mathcal{M}, L_c^2(\mathbb{R}, \frac{dt}{1+t^2}))$ implies that Ψ_* extends to a unique bounded map from $L^1(\mathcal{M}, L_c^2(\mathbb{R}, \frac{dt}{1+t^2}))$ into $L^1(L^\infty(\mathbb{R}) \otimes \mathcal{M}, L_c^2(\widetilde{\Gamma})))$, still denoted by Ψ_*. Consequently, the adjoint $(\Psi_*)^*$ of Ψ_* is bounded from $L^\infty(L^\infty(\mathbb{R}) \otimes \mathcal{M}, L_c^2(\widetilde{\Gamma})))$ to $L^\infty(\mathcal{M}, L_c^2(\mathbb{R}, \frac{dt}{1+t^2}))$ (noting that this adjoint is taken with respect to the anti-dualities). By the very definition of Ψ_*, we have

$$(\Psi_*)^*|_\mathcal{S} = \Psi.$$

This shows that $(\Psi_*)^*$ is an extension of Ψ from $L^\infty(L^\infty(\mathbb{R}) \otimes \mathcal{M}, L_c^2(\widetilde{\Gamma}))$ to $L^\infty(\mathcal{M}, L_c^2(\mathbb{R}, \frac{dt}{1+t^2}))$, which we denote by Ψ again. Being an adjoint, Ψ is w*-continuous.

It remains to show that the so extended map Ψ takes values in $\mathrm{BMO}_c(\mathbb{R}, \mathcal{M})$. Given a bounded interval $I \subset \mathbb{R}$, the w*-topology of $L^\infty(\mathcal{M}, L_c^2(\mathbb{R}, \frac{dt}{1+t^2}))$ induces a topology in $L^\infty(\mathcal{M}, L_c^2(I))$ equivalent to the w*-topology in $L^\infty(\mathcal{M}, L_c^2(I))$. Then by the w*-continuity of Ψ, we deduce that, for every $\varepsilon > 0, I \subset \mathbb{R}, f \in L^1(\mathcal{M}, L_c^2(I))$, there exists a $h \in \mathcal{S}$ such that

$$\tau \int_I f^*(\Psi(g)(t) - \Psi(g)_I) dt$$

$$\leq \tau \int_I f^*(\Psi(h)(t) - \Psi(h)_I) dt + \varepsilon$$

(2.6) $\qquad \leq \|\Psi(h)(t) - \Psi(h)_I\|_{L^\infty(\mathcal{M}, L_c^2(I))} \|f\|_{L^1(\mathcal{M}, L_c^2(I))} + \varepsilon$

and

(2.7) $$\|h\|_{L^\infty(L^\infty(\mathbb{R})\otimes\mathcal{M},L_c^2(\widetilde{\Gamma}))} \leq \|g\|_{L^\infty(L^\infty(\mathbb{R})\otimes\mathcal{M},L_c^2(\widetilde{\Gamma}))} + \varepsilon$$

Combining (2.6), (2.7) and (2.4) we get

$$\int_I f^*(\Psi(g)(t) - \Psi(g)_I)dt$$
$$\leq c|I|\,\|h\|_{L^\infty(L^\infty(\mathbb{R})\otimes\mathcal{M},L_c^2(\widetilde{\Gamma}))}\|f\|_{L^1(\mathcal{M},L_c^2(I))} + \varepsilon$$
$$\leq c|I|(\|g\|_{L^\infty(L^\infty(\mathbb{R})\otimes\mathcal{M},L_c^2(\widetilde{\Gamma}))} + \varepsilon)\|f\|_{L^1(\mathcal{M},L_c^2(I))} + \varepsilon$$

By letting $\varepsilon \to 0$ and taking supremum over all $\|f\|_{L^1(L^\infty(\mathbb{R})\otimes\mathcal{M},L_c^2(\widetilde{\Gamma}))} \leq 1$ and $I \subset \mathbb{R}$, we get $\Psi(g) \in \mathrm{BMO}_c(\mathbb{R},\mathcal{M})$ and

$$\|\Psi(g)\|_{\mathrm{BMO}_c} \leq c\|g\|_{L^\infty(L^\infty(\mathbb{R})\otimes\mathcal{M},L_c^2)}.$$

Therefore, we have extended Ψ to a bounded map from $L^\infty(L^\infty(\mathbb{R})\otimes\mathcal{M},L_c^2(\widetilde{\Gamma}))$ to $\mathrm{BMO}_c(\mathbb{R},\mathcal{M})$, thus completing the proof of the lemma. ∎

Remark. We sketch an alternate proof of the fact that $\varphi = \Psi(h)$ is in $\mathrm{BMO}_c(\mathbb{R},\mathcal{M})$ for $h \in \mathcal{S}$. Let H be the Hilbert space on which $\mathcal{M}$ acts. Recall that $\mathcal{M}_*$ is a quotient space of $B(H)_*$ by the preannihilator of $\mathcal{M}$. Denote the quotient map by q. For every $a,b \in H$, denote $[a \otimes b] = q(a \otimes b)$. Note that $\tau(m^*[a \otimes b]) = \tau([m^*(a \otimes b)]) = \langle m(b), \overline{a}\rangle, \forall m \in \mathcal{M}$. From (1.16) and the classical duality between $\mathrm{BMO}(\mathbb{R},H)$ and $H^1(\mathbb{R},H)$,

$$\|\varphi\|_{\mathrm{BMO}_c(\mathbb{R},\mathcal{M})} = \sup_{e \in H, \|e\|_H = 1} \|\varphi e\|_{\mathrm{BMO}(\mathbb{R},H)}.$$

$$\leq c \sup_{e \in H, \|e\|_H = 1} \sup_{\|g\|_{H^1(\mathbb{R},H)} = 1} \left|\int_{-\infty}^{+\infty} \langle \varphi(e), \overline{g}\rangle\, dt\right|$$

(2.8) $$= c \sup_{e \in H, \|e\|_H = 1} \sup_{\|g\|_{H^1(\mathbb{R},H)} = 1} \left|\tau \int_{-\infty}^{+\infty} \varphi^*[g \otimes e]dt\right|$$

Let $f = [g \otimes e]$. Noting that

$$|\nabla f|^2 = \langle \nabla g, \nabla g\rangle [e \otimes e] = |\nabla g|^2 [e \otimes e],$$

we get

(2.9) $$\tau(S_c(f)(t)) = \Big(\iint_\Gamma |\nabla g(t+x,y)|^2\, dxdy\Big)^{\frac{1}{2}}.$$

Thus $\|f\|_{\mathcal{H}_c^1(\mathbb{R},\mathcal{M})} = 1$ if $\|g\|_{H^1(\mathbb{R},H)} = 1$ and $\|e\|_H = 1$. Therefore

$$\|\varphi\|_{\mathrm{BMO}_c(\mathbb{R},\mathcal{M})} \leq c \sup_{\|f\|_{\mathcal{H}_c^1(\mathbb{R},\mathcal{M})}=1} \left|\tau\int_{-\infty}^{+\infty} \varphi^* f dt\right|$$

$$= c \sup_{\|f\|_{\mathcal{H}_c^1(\mathbb{R},\mathcal{M})}=1} \left|\tau \int_{-\infty}^{+\infty}\iint_\Gamma h^*(x,y,t)\nabla f(t+x,y)dydxdt\right|$$

$$\leq c\|h\|_{L^\infty(L^\infty(\mathbb{R})\otimes\mathcal{M},L_c^2)}.$$

COROLLARY 2.3. *Let* $f \in L^1(\mathcal{M}, L_c^2(\mathbb{R}, (1+s^2)ds))$ *with* $\int f ds = 0$. *Then* $f \in \mathcal{H}_c^1(\mathbb{R}, \mathcal{M})$ *and*
$$\|f\|_{\mathcal{H}_c^1} \leq c \|f\|_{L^1(\mathcal{M}, L_c^2(\mathbb{R}, (1+s^2)ds))}$$

Proof. By Lemma 2.2, the assumption that $\int f ds = 0$ and Proposition 1.3, we have

$$\begin{aligned}
\|f\|_{\mathcal{H}_c^1} &= \|\nabla f(t+x,y)\chi_\Gamma\|_{L^1(L^\infty(\mathbb{R})\bar{\otimes}\mathcal{M}, L_c^2)} \\
&= \sup_{\|h\|_{L^\infty(L^\infty(\mathbb{R})\bar{\otimes}\mathcal{M}, L_c^2)} \leq 1} \left|\tau \int \iint_\Gamma h^* \nabla f(t+x,y)dxdydt\right| \\
&= \sup_{\|h\|_{L^\infty(L^\infty(\mathbb{R})\bar{\otimes}\mathcal{M}, L_c^2)} \leq 1} \left|\tau \int_\mathbb{R} (\Psi(h))^*(s)f(s)ds\right| \\
&\leq c \sup_{\|\varphi\|_{\text{BMO}_c(\mathbb{R},\mathcal{M})} \leq 1} \left|\tau \int_\mathbb{R} \varphi^*(s)f(s)ds\right| \\
&\leq c \sup_{\|\varphi\|_{L^\infty(\mathcal{M}, L_c^2(\mathbb{R}, \frac{ds}{1+s^2}))} \leq 1} \left|\tau \int_\mathbb{R} \varphi^*(s)(1+s^2)f(s)\frac{ds}{1+s^2}\right| \\
&\leq c \|(1+s^2)f(s)\|_{L^1(\mathcal{M}, L_c^2(\mathbb{R}, \frac{ds}{1+s^2}))} \\
&= c \|f\|_{L^1(\mathcal{M}, L_c^2(\mathbb{R}, (1+s^2)ds))} \cdot \blacksquare
\end{aligned}$$

Remark. In particular, every $S_\mathcal{M}$-valued simple function f with $\int f ds = 0$ is in $\mathcal{H}_c^1(\mathbb{R}, \mathcal{M})$. Consequently, by the remark before Proposition 1.3, $\mathcal{H}_c^1(\mathbb{R}, \mathcal{M}) \cap \mathcal{H}_c^p(\mathbb{R}, \mathcal{M})$ is dense in $\mathcal{H}_c^p(\mathbb{R}, \mathcal{M})$ ($p > 1$).

2. The duality theorem of operator valued $\mathcal{H}^1$ and BMO

Denote by $\mathcal{H}_{c0}^1(\mathbb{R}, \mathcal{M})$ (resp. $\mathcal{H}_{r0}^1(\mathbb{R}, \mathcal{M})$) the family of functions f in $\mathcal{H}_c^1(\mathbb{R}, \mathcal{M})$ (resp. $\mathcal{H}_r^1(\mathbb{R}, \mathcal{M}), \mathcal{H}_{cr}^1(\mathbb{R}, \mathcal{M})$) such that $f \in L^1(\mathcal{M}, L_c^2(\mathbb{R}, (1+t^2)dt))$ (resp. $L^1(\mathcal{M}, L_r^2(\mathbb{R}, (1+t^2)dt))$. It is easy to see that $\mathcal{H}_{c0}^1(\mathbb{R}, \mathcal{M})$ (resp. $\mathcal{H}_{r0}^1(\mathbb{R}, \mathcal{M})$) is a dense subspace of $\mathcal{H}_c^1(\mathbb{R}, \mathcal{M})$ (resp. $\mathcal{H}_r^1(\mathbb{R}, \mathcal{M})$)). Let
$$\mathcal{H}_{cr0}^1(\mathbb{R}, \mathcal{M}) = \mathcal{H}_{c0}^1(\mathbb{R}, \mathcal{M}) + \mathcal{H}_{r0}^1(\mathbb{R}, \mathcal{M}).$$
Then $\mathcal{H}_{cr0}^1(\mathbb{R}, \mathcal{M})$ is a dense subspace of $\mathcal{H}_{cr}^1(\mathbb{R}, \mathcal{M})$. Recall that we have proved in Chapter 1 that $\text{BMO}_c(\mathbb{R}, \mathcal{M}) \subseteq L^\infty(\mathcal{M}, L_c^2(\mathbb{R}, \frac{dt}{1+t^2}))$. Thus by Proposition 1.1 $\langle \varphi, f \rangle = \int_{-\infty}^{+\infty} \varphi^* f dt$ exists in $L^1(\mathcal{M})$ for all $\varphi \in \text{BMO}_c(\mathbb{R}, \mathcal{M})$ and $f \in \mathcal{H}_{c0}^1(\mathbb{R}, \mathcal{M})$ (see our convention after Proposition 1.1).

THEOREM 2.4. (a) *We have* $(\mathcal{H}_c^1(\mathbb{R}, \mathcal{M}))^* = \text{BMO}_c(\mathbb{R}, \mathcal{M})$ *with equivalent norms. More precisely, every* $\varphi \in \text{BMO}_c(\mathcal{M})$ *defines a continuous linear functional on* $\mathcal{H}_c^1(\mathbb{R}, \mathcal{M})$ *by*

$$(2.10) \qquad l\varphi(f) = \tau \int_{-\infty}^{+\infty} \varphi^* f dt; \qquad \forall f \in \mathcal{H}_{c0}^1(\mathbb{R}, \mathcal{M}).$$

Conversely, every $l \in (\mathcal{H}_c^1(\mathbb{R}, \mathcal{M}))^*$ *can be given as above by some* $\varphi \in \text{BMO}_c(\mathbb{R}, \mathcal{M})$. *Moreover, there exists a universal constant* $c > 0$ *such that*

$$c^{-1}\|\varphi\|_{\text{BMO}_c} \leq \|l\varphi\|_{(\mathcal{H}_c^1)^*} \leq c\|\varphi\|_{\text{BMO}_c}.$$

Thus $(\mathcal{H}_c^1(\mathbb{R}, \mathcal{M}))^* = \mathrm{BMO}_c(\mathbb{R}, \mathcal{M})$ with equivalent norms.

(b) Similarly, $(\mathcal{H}_r^1(\mathbb{R}, \mathcal{M}))^* = \mathrm{BMO}_r(\mathbb{R}, \mathcal{M})$ with equivalent norms.

(c) $(\mathcal{H}_{cr}^1(\mathbb{R}, \mathcal{M}))^* = \mathrm{BMO}_{cr}(\mathbb{R}, \mathcal{M})$ with equivalent norms.

Our proof of Theorem 2.4 requires two technical variants of the square functions $G_c(f)$ and $S_c(f)$. These are operator valued functions defined as follows:

$$(2.11) \qquad G_c(f)(x, y) = (\int_y^\infty |\nabla f(x, s)|^2 s ds)^{\frac{1}{2}},$$

$$(2.12) \qquad S_c(f)(x, y) = (\iint_{\Gamma(0,y)} |\nabla f(t+x, s)|^2 dt ds)^{\frac{1}{2}}$$

where $y \geq 0, \Gamma(0, y) = \{(t, s) : |t| < s - y, s \geq y\}$ and f is $S_\mathcal{M}$-valued simple function. Note that $G_c(f)(x, 0)$ and $S_c(f)(x, 0)$ are just $G_c(f)$ and $S_c(f)$ defined in Chapter 1.

LEMMA 2.5.
$$G_c(f)(x, y) \leq 2\sqrt{2} S_c(f)(x, \frac{y}{2}) .$$

Proof. It suffices to prove this inequality for $x = 0$. Let us denote by B_s the ball centered at $(0, s)$ and tangent to the boundary of $\Gamma(0, \frac{y}{2}), \forall s > y$. By the harmonicity of ∇f, we get

$$\nabla f(0, s) = \frac{2}{\pi(s - \frac{y}{2})^2} \int_{B_s} \nabla f(x, u) dx du$$

By (1.12),

$$|\nabla f(0, s)|^2 \leq \frac{8}{\pi s^2} \int_{B_s} |\nabla f(x, u)|^2 dx du$$

Integrating this inequality, we obtain

$$(2.13) \qquad \int_y^\infty s|\nabla f(0, s)|^2 ds \leq \int_y^\infty \frac{8}{\pi s} \int_{B_s} |\nabla f(x, u)|^2 dx du ds$$

However $(x, u) \in B_s$ clearly implies that $\frac{u}{2} \leq s \leq 4u$. Thus, the right hand side of (2.13) is majorized by

$$\int_{\Gamma(0, \frac{y}{2})} |\nabla f(x, u)|^2 \int_{\frac{u}{2}}^{4u} \frac{8}{\pi s} ds dx du \leq 8 S_c^2(f)(0, \frac{y}{2})$$

Therefore $G_c(f)(0, y) \leq 2\sqrt{2} S_c(f)(0, \frac{y}{2})$. ∎

Proof of Theorem 2.4. (i) We will first prove

$$(2.14) \qquad |l_\varphi(f)| \leq c \|\varphi\|_{\mathrm{BMO}_c} \|f\|_{\mathcal{H}_c^1}$$

when both f and φ have compact support. Once this is done, by Lemma 1.5, we can see (2.14) holds for any $\varphi \in \mathrm{BMO}_c(\mathbb{R}, \mathcal{M})$ and any compactly supported $f \in \mathcal{H}_{c0}^1(\mathbb{R}, \mathcal{M})$. Then recall that by Proposition 1.3

$$\mathrm{BMO}_c(\mathbb{R}, \mathcal{M}) \subset L^\infty(\mathcal{M}, L_c^2(\mathbb{R}, \frac{dt}{1+t^2}))$$

and by Corollary 2.3

$$\|f\|_{\mathcal{H}_c^1} \leq c \|f\|_{L^1(\mathcal{M}, L_c^2(\mathbb{R}, (1+t^2) dt))} , \quad \forall f \in \mathcal{H}_{c0}^1(\mathbb{R}, \mathcal{M}),$$

we deduce (2.14) for all $\varphi \in \mathrm{BMO}_c(\mathbb{R}, \mathcal{M})$, $f \in \mathcal{H}^1_{c0}(\mathbb{R}, \mathcal{M})$ by choosing compactly supported $f_n \in \mathcal{H}^1_{c0}(\mathbb{R}, \mathcal{M}) \to f$ in $L^1(\mathcal{M}, L^2_c(\mathbb{R}, (1+t^2)dt))$. Finally, from the density of $\mathcal{H}^1_{c0}(\mathbb{R}, \mathcal{M})$ in $\mathcal{H}^1_c(\mathbb{R}, \mathcal{M})$, l_φ defined in (2.10) extends to a continuous functional on $\mathcal{H}^1_c(\mathbb{R}, \mathcal{M})$.

Let us now prove (2.14) for compactly supported $f \in \mathcal{H}^1_{c0}(\mathbb{R}, \mathcal{M})$ and compactly supported $\varphi \in \mathrm{BMO}_c(\mathbb{R}, \mathcal{M})$. By approximation we may assume that τ is finite and $G_c(f)(x,y)$ is invertible in $\mathcal{M}$ for every $(x,y) \in \mathbb{R}^2_+$. Recall that $\triangle(\varphi^* f) = 2\nabla \varphi^* \nabla f$. By Green's theorem and the Cauchy-Schwarz inequality

$$
\begin{aligned}
|l\varphi(f)| &= 2|\tau \iint_{\mathbb{R}^2_+} \nabla\varphi^* \nabla f y dy dx| \\
&\leq 2(\tau \iint_{\mathbb{R}^2_+} G_c^{-\frac{1}{2}}(f)|\nabla f|^2 G_c^{-\frac{1}{2}}(f) y dy dx)^{\frac{1}{2}} (\tau \iint_{\mathbb{R}^2_+} G_c^{\frac{1}{2}}(f)|\nabla \varphi|^2 G_c^{\frac{1}{2}}(f) y dy dx)^{\frac{1}{2}} \\
&= 2(\tau \iint_{\mathbb{R}^2_+} G_c^{-1}(f)|\nabla f|^2 y dy dx)^{\frac{1}{2}} (\tau \iint_{\mathbb{R}^2_+} G_c(f)|\nabla \varphi|^2 y dy dx)^{\frac{1}{2}} \\
&= 2I \bullet II,
\end{aligned}
$$

Note here $G_c(f)$ is the function of two variables defined by (2.11), which is differentiable in the weak-* sense. For I we have

$$
\begin{aligned}
I^2 &= \tau \int_{-\infty}^{+\infty} \int_0^\infty -G_c^{-1}(f) \frac{\partial G_c^2(f)}{\partial y} dy dx \\
&= \tau \int_{-\infty}^{+\infty} \int_0^\infty (-G_c^{-1}(f) \frac{\partial G_c(f)}{\partial y} G_c(f) - \frac{\partial G_c(f)}{\partial y}) dy dx \\
&= 2\tau \int_{-\infty}^{+\infty} \int_0^\infty -\frac{\partial G_c(f)}{\partial y} dy dx \\
&= 2\tau \int_{-\infty}^{+\infty} G_c(f)(x, 0) dx \\
&\leq 4\sqrt{2}\tau \int_{-\infty}^{+\infty} S_c(f)(x, 0) dx \\
&= 4\sqrt{2} \|f\|_{\mathcal{H}^1_c}.
\end{aligned}
$$

To estimate II, we create a square net partition in $\mathbb{R}^2_+$ as follows:

$$\sigma(i,j) = \{(x,y) : (i-1)2^j < x \leq i 2^j, 2^j \leq y < 2^{j+1}\}, \quad \forall i, j \in \mathbb{Z}.$$

Let $C_{i,j}$ denote the center of $\sigma(i,j)$. Define

$$
\begin{aligned}
\widetilde{S}_c(f)(x,y) &= S_c(f)(C_{i,j}), \quad \forall (x,y) \in \sigma(i,j), \\
d_k(x) &= \widetilde{S}_c(f)(x, 2^k) - \widetilde{S}_c(f)(x, 2^{k+1}), \quad \forall x \in \mathbb{R}.
\end{aligned}
$$

It is easy to check that

$$S_c(f)(x,2y) \leq \widetilde{S}_c(f)(x,y) \leq S_c(f)(x,\frac{y}{2}),$$
$$d_k(x) \geq 0, \quad \forall x \in \mathbb{R},$$
$$\widetilde{S}_c(f)(x,y) = \sum_{k=j}^{\infty} d_k(x), \quad \forall 2^j \leq y < 2^{j+1},$$

(2.15) $$S_c(f)(x,0) = \sum_{k=-\infty}^{\infty} d_k(x).$$

Now by Lemma 2.5 and (2.15)

$$\begin{aligned}
II^2 &= \tau \int_{-\infty}^{+\infty} \int_0^{\infty} G_c(f)(x,y) |\nabla \varphi|^2 y dy dx \\
&\leq 2\sqrt{2}\tau \int_{-\infty}^{+\infty} \int_0^{\infty} \widetilde{S}_c(f)(x,\frac{y}{4}) |\nabla \varphi|^2 y dy dx \\
&= 2\sqrt{2}\tau \int_{-\infty}^{+\infty} \sum_{k=-\infty}^{\infty} \widetilde{S}_c(f)(x,2^k) \int_{2^{k+2}}^{2^{k+3}} |\nabla \varphi|^2 y dy dx \\
&= 2\sqrt{2}\tau \int_{-\infty}^{+\infty} \sum_{k=-\infty}^{\infty} (\sum_{j=k}^{\infty} d_j(x)) \int_{2^{k+2}}^{2^{k+3}} |\nabla \varphi|^2 y dy dx \\
&= 2\sqrt{2}\tau \int_{-\infty}^{+\infty} \sum_{j=-\infty}^{\infty} d_j(x) \int_0^{2^{j+3}} |\nabla \varphi|^2 y dy dx \\
&= 2\sqrt{2}\tau \sum_{i=-\infty}^{\infty} \sum_{j=-\infty}^{\infty} d_j(i2^j) \int_{(i-1)2^j}^{i2^j} \int_0^{2^{j+3}} |\nabla \varphi|^2 y dy dx
\end{aligned}$$

Hence by Lemma 1.4

$$\begin{aligned}
II^2 &\leq c\tau \sum_{i=-\infty}^{\infty} \sum_{j=-\infty}^{\infty} d_j(i2^j) 2^j \|\varphi\|_{\mathrm{BMO}_c}^2 \\
&= c\|\varphi\|_{\mathrm{BMO}_c}^2 \tau \sum_{j=-\infty}^{\infty} \int_{-\infty}^{+\infty} d_j(x) dx \\
&= c\|\varphi\|_{\mathrm{BMO}_c}^2 \tau \int_{-\infty}^{+\infty} S_c(f)(x,0) dx \\
&= c\|\varphi\|_{\mathrm{BMO}_c}^2 \|f\|_{\mathcal{H}_c^1}.
\end{aligned}$$

Combining the preceding estimates on I and II, we get

$$|l\varphi(f)| \leq c\|\varphi\|_{\mathrm{BMO}_c} \|f\|_{\mathcal{H}_c^1}.$$

Therefore, $l\varphi$ defines a continuous functional on $\mathcal{H}_c^1$ of norm smaller than $c\|\varphi\|_{\mathrm{BMO}_c}$.

(ii) Now suppose $l \in (\mathcal{H}_c^1(\mathbb{R},\mathcal{M}))^*$. Then by the Hahn-Banach theorem l extends to a continuous functional on $L^1(L^{\infty}(\mathbb{R}) \otimes \mathcal{M}, L_c^2(\widetilde{\Gamma}))$ of the same norm. Thus by

$$(L^1(L^{\infty}(\mathbb{R}) \otimes \mathcal{M}, L_c^2(\widetilde{\Gamma})))^* = L^{\infty}(L^{\infty}(\mathbb{R}) \otimes \mathcal{M}, L_c^2(\widetilde{\Gamma}))$$

there exists $g \in L^\infty(L^\infty(\mathbb{R}) \otimes \mathcal{M}, L_c^2(\widetilde{\Gamma}))$ such that

$$||g||^2_{L^\infty(L^\infty(\mathbb{R}) \otimes \mathcal{M}, L_c^2(\widetilde{\Gamma}))} = \sup_{t \in \mathbb{R}} || \iint_\Gamma g^*(x,y,t) g(x,y,t) dy dx ||_{L^\infty(\mathbb{R}) \otimes \mathcal{M}} = ||l||^2$$

and

$$l(f) = \tau \int_{-\infty}^{+\infty} \iint_\Gamma g^*(x,y,t) \nabla f(t+x, y) dy dx dt, \ \forall \ f \in \mathcal{H}^1_{c0}(\mathbb{R}, \mathcal{M}).$$

Let $\varphi = \Psi(g)$, where Ψ is the extension given by Lemma 2.2. By that lemma $\varphi \in \text{BMO}_c(\mathbb{R}, \mathcal{M})$ and

$$||\varphi||_{\text{BMO}_c} \le c ||g||_{L^\infty(L^\infty(\mathbb{R}) \otimes \mathcal{M}, L_c^2(\widetilde{\Gamma}))} = c||l||.$$

Then we must show that

$$l(f) = \tau \int_{-\infty}^{+\infty} \varphi^*(s) f(s) ds, \ \forall \ f \in \mathcal{H}^1_{c0}(\mathbb{R}, \mathcal{M}).$$

But this follows from the second part of the proof of Lemma 2.2 in virtue of the w*-continuity of Ψ. Therefore, we have accomplished the proof of the theorem concerning $\mathcal{H}^1_c(\mathbb{R}, \mathcal{M})$ and $\text{BMO}_c(\mathbb{R}, \mathcal{M})$. Passing to adjoints yields the part on $\mathcal{H}^1_r(\mathbb{R}, \mathcal{M})$ and BMO_r. Finally, the duality between $\mathcal{H}^1_{cr}(\mathbb{R}, \mathcal{M})$ and $\text{BMO}_{cr}(\mathbb{R}, \mathcal{M})$ is obtained from the classical fact that the dual of a sum is the intersection of the duals. ∎

COROLLARY 2.6. *$\varphi \in \text{BMO}_c(\mathbb{R}, \mathcal{M})$ if and only if $d\lambda\varphi = |\nabla \varphi|^2 y dx dy$ is an $\mathcal{M}$-valued Carleson measure on $\mathbb{R}^2_+$, and $c^{-1} N(\lambda_\varphi) \le ||\varphi||^2_{\text{BMO}_c} \le c N(\lambda_\varphi)$.*

Proof. From the first part of the proof of Theorem 2.4, if φ is such that $d\lambda_\varphi = |\nabla \varphi|^2 y dx dy$ is an $\mathcal{M}$-valued Carleson measure, then φ defines a continuous linear functional $l_\varphi = \tau \int_{-\infty}^{+\infty} \varphi^* f dt$ on $\mathcal{H}^1_{c0}(\mathbb{R}, \mathcal{M})$ and

$$||l_\varphi||_{(\mathcal{H}^1_c)^*} \le c N^{\frac{1}{2}}(\lambda_\varphi)$$

Therefore by Theorem 2.4 again there exists a function $\varphi' \in \text{BMO}_c(\mathbb{R}, \mathcal{M})$ with $||\varphi'||^2_{\text{BMO}_c} \le ||l_\varphi||^2_{(\mathcal{H}^1_c)^*} \le c N(\lambda_\varphi)$ such that

$$\tau \int_{-\infty}^{+\infty} \varphi^* f dt = \tau \int_{-\infty}^{+\infty} \varphi'^* f dt.$$

Thus $\varphi = \varphi'$ and $\varphi \in \text{BMO}_c(\mathbb{R}, \mathcal{M})$ with $||\varphi||^2_{\text{BMO}_c} \le c N(\lambda_\varphi)$. The converse had been already proved in Lemma 1.4. ∎

COROLLARY 2.7. *For $f \in \mathcal{H}^1_c(\mathbb{R}, \mathcal{M})$, we have*

$$c^{-1} ||G_c(f)||_1 \le ||S_c(f)||_1 \le c ||G_c(f)||_1$$

Proof. By Theorem 2.4 and the first part of its proof, we have

$$||S_c(f)||_1 = ||f||_{\mathcal{H}^1_c} \le c \sup_{||\varphi||_{\text{BMO}_c}=1} \left| \tau \int f \varphi^* dt \right| \le c ||G_c(f)||_1^{\frac{1}{2}} ||S_c(f)||_1^{\frac{1}{2}}$$

Therefore

$$||S_c(f)||_1 \le c ||G_c(f)||_1$$

The converse is an immediate consequence of Lemma 2.5. ∎

Remark. The technique used in the proof of Lemma 2.5 is classical (see [**33**]). The method to prove Theorem 2.4 is inspired by the analogous one for martingales (see [**7**], [**10**], [**28**]).

3. The atomic decomposition of operator valued $\mathcal{H}^1$

As in the classical case, the duality between $\mathcal{H}_c^1(\mathbb{R}, \mathcal{M})$ and $\mathrm{BMO}_c(\mathbb{R}, \mathcal{M})$ implies an atomic decomposition of $\mathcal{H}_c^1(\mathbb{R}, \mathcal{M})$. The rest of this chapter is devoted to this atomic decomposition. We say that a function $a \in L^1(\mathcal{M}, L_c^2(\mathbb{R}))$ is an $\mathcal{M}_c$-atom if
(i) a is supported in a bounded interval I;
(ii) $\int_I a\, dt = 0$;
(iii) $\tau(\int_I |a|^2 dt)^{\frac{1}{2}} \leq |I|^{-\frac{1}{2}}$.

Let $\mathcal{H}_c^{1,at}(\mathbb{R}, \mathcal{M})$ be the space of all f which admit a representation of the form
$$f = \sum_{i \in \mathbb{N}} \lambda_i a_i,$$
where the a_i's are $\mathcal{M}_c$-atoms and $\lambda_i \in \mathbb{C}$ are such that $\sum_{i \in \mathbb{N}} |\lambda_i| < \infty$. We equip $\mathcal{H}_c^{1,at}(\mathbb{R}, \mathcal{M})$ with the following norm
$$\|f\|_{\mathcal{H}_c^{1,at}} = \inf\{\sum_{i \in \mathbb{N}} |\lambda_i|; f = \sum_{i \in \mathbb{N}} \lambda_i a_i; a_i \text{ are } \mathcal{M}_c\text{-atoms}, \lambda_i \in \mathbb{C}\}$$

Similarly, we define $\mathcal{H}_r^{1,at}(\mathbb{R}, \mathcal{M})$. Then we set
$$\mathcal{H}_{cr}^{1,at}(\mathbb{R}, \mathcal{M}) = \mathcal{H}_c^{1,at}(\mathbb{R}, \mathcal{M}) + \mathcal{H}_r^{1,at}(\mathbb{R}, \mathcal{M}).$$

THEOREM 2.8. $\mathcal{H}_c^{1,at}(\mathbb{R}, \mathcal{M}) = \mathcal{H}_c^1(\mathbb{R}, \mathcal{M})$ *with equivalent norms.*

Proof. It is enough to prove $(\mathcal{H}_c^{1,at}(\mathbb{R}, \mathcal{M}))^* = \mathrm{BMO}_c(\mathbb{R}, \mathcal{M})$. Now, for any $\varphi \in \mathrm{BMO}_c(\mathbb{R}, \mathcal{M})$ and $f \in \mathcal{H}_c^{1,at}(\mathbb{R}, \mathcal{M})$ with $f = \sum_{i \in \mathbb{N}} \lambda_i a_i$ as above, by the Cauchy-Schwarz inequality we have

$$\begin{aligned}
|\tau \int \varphi^* f\, dt| &\leq \sum_{i \in \mathbb{N}} |\lambda_i \tau \int_{I_i} (\varphi - \varphi_{I_i})^* a_i\, dt| \\
&\leq \sum_{i \in \mathbb{N}} |\lambda_i| \tau (\int_{I_i} |a_i|^2 dt)^{\frac{1}{2}} \left\| (\int_{I_i} |\varphi - \varphi_{I_i}|^2 dt)^{\frac{1}{2}} \right\|_{\mathcal{M}} \\
&\leq \|\varphi\|_{\mathrm{BMO}_c} \sum_{i \in \mathbb{N}} |\lambda_i|.
\end{aligned}$$

Thus $\mathrm{BMO}_c(\mathbb{R}, \mathcal{M}) \subset (\mathcal{H}_c^{1,at}(\mathbb{R}, \mathcal{M}))^*$ (a contractive inclusion). To prove the converse inclusion, we denote by $L_0^1(\mathcal{M}, L_c^2(I))$ the space of functions $f \in L^1(\mathcal{M}, L_c^2(I))$ with $\int f\, dt = 0$. Notice that $L_0^1(\mathcal{M}, L_c^2(I)) \in \mathcal{H}_c^{1,at}(\mathbb{R}, \mathcal{M})$ for every bounded I. Thus, every continuous functional l on $\mathcal{H}_c^{1,at}(\mathbb{R}, \mathcal{M})$ induces a continuous functional on $L_0^1(\mathcal{M}, L_c^2(I))$ with norm smaller than $|I|^{\frac{1}{2}} \|l\|_{(\mathcal{H}_c^{1,at})^*}$. Consequently, we can choose a sequence $(\varphi_n)_{n \geq 1}$ satisfying the following conditions:

$$\begin{aligned}
&l(a) = \tau \int \varphi_n^* a\, dt, \quad \forall \mathcal{M}_c\text{- atom } a \text{ with } \mathrm{supp}\, a \subset (-n, n], \\
&\|\varphi_n\|_{L^\infty(\mathcal{M}, L_c^2((-n,n]))} \leq c\sqrt{n}\, \|l\|_{(\mathcal{H}_c^{1,at})^*}; \\
&\varphi_n|_{(-m,m]} = \varphi_m, \quad \forall n > m.
\end{aligned}$$

Let $\varphi(t) = \varphi_n(t), \forall t \in (-n, -n+1] \cup (n-1, n], n > 0$. We then have $\varphi \in L^\infty(\mathcal{M}, L_c^2(\mathbb{R}, \frac{dt}{1+t^2}))$ and

$$l(a) = \tau \int \varphi^* a \, dt, \quad \forall \mathcal{M}_c\text{- atom } a.$$

Considering $[g \otimes e]$ as defined in the remark after Lemma 2.2, by (2.8) and (2.9) we have

$$\begin{aligned}
\|\varphi\|_{\mathrm{BMO}_c} &\leq c \sup_{e \in H, \|e\|_H = 1} \sup_{\|g\|_{H^1(\mathbb{R},H)}=1} \left| \tau \int_{-\infty}^{+\infty} \varphi^*[g \otimes e] dt \right| \\
&\leq \sup_{\|f\|_{\mathcal{H}_c^{1,at}}=1} \left| \tau \int_{-\infty}^{+\infty} \varphi^* f \, dt \right| \\
&= \|l\|_{(\mathcal{H}_c^{1,at})^*}. \quad \blacksquare
\end{aligned}$$

COROLLARY 2.9. $\mathcal{H}_r^{1,at}(\mathbb{R}, \mathcal{M}) = \mathcal{H}_r^1(\mathbb{R}, \mathcal{M})$ and $\mathcal{H}_{cr}^{1,at}(\mathbb{R}, \mathcal{M}) = \mathcal{H}_{cr}^1(\mathbb{R}, \mathcal{M})$ with equivalent norms.

Remark. The $\mathcal{M}$-atom considered in this section is a noncommutative analogue of the classical 2-atom for H^1 space. It seems difficult to consider the noncommutative analogues of the classical p–atom for $p \neq 2$.

Remark. We only considered the functions defined on $\mathbb{R}$ in this chapter. However, one can check that all the proofs work well for the functions defined on $\mathbb{R}^n$. And the analogous results can be proved similarly for the functions defined on $\mathbb{T}^n$, where $\mathbb{T}$ is the unit circle. Moreover, the relevant constants are independent of n.

CHAPTER 3

The Maximal Inequality

1. The noncommutative Hardy-Littlewood maximal inequality

We recall the definition of the noncommutative maximal norm introduced by Pisier (see [**27**]) and Junge (see [**14**]). Let $0 < p \leq \infty$, and let $(a_n)_{n \in \mathbb{Z}}$ be a sequence of elements in $L^p(\mathcal{M})$. Set

$$(3.1) \quad \left\| \sup_{n \in \mathbb{Z}} |a_n| \right\|_{L^p(\mathcal{M})} = \inf_{a_n = a y_n b} \|a\|_{L^{2p}(\mathcal{M})} \|b\|_{L^{2p}(\mathcal{M})} \sup_n \|y_n\|_{\mathcal{M}}$$

where the infimum is taken over all $a, b \in L_{2p}(\mathcal{M})$ and all bounded sequences $(y_n)_{n \in \mathbb{Z}} \in \mathcal{M}$ such that $a_n = a y_n b$. By convention, if $(a_n)_{n \in \mathbb{Z}}$ does not have such a representation, we define $\|\sup_{n \in \mathbb{Z}} |a_n|\|_{L^p(\mathcal{M})}$ as $+\infty$.

If $p \geq 1$ and $(a_n)_{n \in \mathbb{Z}}$ is a sequence of positive elements, it was proved by Junge and Xu (see [**14**], Remark 3.7; [**17**], Proposition 2.1) that (with q the index conjugate to p)

$$(3.2) \quad \left\| \sup_{n \in \mathbb{Z}} |a_n| \right\|_{L^p(\mathcal{M})} = \sup \left\{ \sum_{n \in \mathbb{Z}} \tau(a_n b_n) : b_n \in L^q(\mathcal{M}), b_n \geq 0, \left\| \sum_{n \in \mathbb{Z}} b_n \right\|_{L^q(\mathcal{M})} \leq 1 \right\}.$$

In this case, $\|\sup_{n \in \mathbb{Z}} |a_n|\|_{L^p(\mathcal{M})} < \infty$ if and only if there exists $a \in L^p(\mathcal{M}), a > 0$ and a sequence of positive contractions y_n such that $a_n = a^{\frac{1}{2}} y_n a^{\frac{1}{2}}$, $\forall n \in \mathbb{Z}$, and moreover,

$$\left\| \sup_n |a_n| \right\|_{L^p(\mathcal{M})} = \inf\{\|a\|_{L^p(\mathcal{M})} : a > 0, a_n \leq a, \forall n \in \mathbb{Z}\}.$$

We define similarly $\|\sup_{\lambda \in \Lambda} |a(\lambda)|\|_p$ if Λ is a countable set. If Λ is uncountable we set

$$(3.3) \quad \left\| \sup_{\lambda \in \Lambda} |a(\lambda)| \right\|_{L^p(\mathcal{M})} = \sup_{(\lambda_n)_{n \in \mathbb{Z}} \in \Lambda} \left\| \sup_{n \in \mathbb{Z}} |a(\lambda_n)| \right\|_{L^p(\mathcal{M})}.$$

Please note that $\sup_\lambda |a(\lambda)|$ does not make any sense in the noncommutative setting and $\|\sup_{\lambda \in \Lambda} |a(\lambda)|\|_{L^p(\mathcal{M})}$ is just a notation. Also note that

$$(3.4) \quad \left\| \sup_{\lambda \in \Lambda} |a(\lambda)| \right\|_{L^\infty(\mathcal{M})} = \sup_{\lambda \in \Lambda} \|a(\lambda)\|_{L^\infty(\mathcal{M})}.$$

and for $1 \leq p \leq \infty$,

$$(3.5) \quad \left\| \sup_{\lambda \in \Lambda} |a(\lambda)| \right\|_{L^p(\mathcal{M})} = \sup_{J \subset \Lambda \text{ finite}} \left\| \sup_{n \in J} |a(\lambda_n)| \right\|_{L^p(\mathcal{M})}.$$

1. THE NONCOMMUTATIVE HARDY-LITTLEWOOD MAXIMAL INEQUALITY

The main result of this chapter is the noncommutative Hardy-Littlewood maximal inequality. We will reduce it to the noncommutative Doob maximal inequality for martingales already established by M. Junge [9]. To this end, we need to introduce two increasing filtration of dyadic σ-algebras on $\mathbb{R}$. The key property of these σ-algebras is that any interval of $\mathbb{R}$ is contained in an atom belonging to one of these σ-algebras with a comparable size (see Proposition 3.1 below). This approach is very simple. And we will need it later when prove $\mathrm{BMO}_c(\mathbb{R}, \mathcal{M})$ is the intersection of two dyadic BMO spaces. That is one of the reasons that we do not follow the classical ways to dominate Hardy-Littlewood maximal functions by the correspondent dyadic ones.

The two increasing filtrations of dyadic σ-algebras $\mathcal{D} = \{\mathcal{D}_n\}_{n \in \mathbb{Z}}, \mathcal{D}' = \{\mathcal{D}'_n\}_{n \in \mathbb{Z}}$ that we will need are defined as follows: The first one, $\mathcal{D} = \{\mathcal{D}_n\}_{n \in \mathbb{Z}}$, is simply the usual dyadic filtration, that is, $\mathcal{D}_n$ is the σ-algebra generated by the atoms

$$D_n^k = (k2^{-n}, (k+1)2^{-n}]; \quad k \in \mathbb{Z}.$$

The definition of $\mathcal{D}' = \{\mathcal{D}'_n\}_{n \in \mathbb{Z}}$ is a little more complicated. For an even integer n, the atoms of $\mathcal{D}'_n$ are given by

$$D'^k_n = ((k+\frac{1}{3})2^{-n}, (k+\frac{4}{3})2^{-n}], \quad k \in \mathbb{Z};$$

while for an odd integer n, $\mathcal{D}'_n$ is generated by the atoms

$$D'^k_n = ((k+\frac{2}{3})2^{-n}, (k+\frac{5}{3})2^{-n}], \quad k \in \mathbb{Z}.$$

It is easy to see that $\mathcal{D}' = \{\mathcal{D}'_n\}_{n \in \mathbb{Z}}$ is indeed an increasing filtration.

The following simple observation is the key of our approach.

PROPOSITION 3.1. *For any interval $I \subset \mathbb{R}$, there exist $k_I, N \in \mathbb{Z}$ such that $I \subset D_N^{k_I}$ and $|D_N^{k_I}| \leq 6|I|$ or $I \subset D'^{k_I}_N$ and $|D'^{k_I}_N| \leq 6|I|$, the constant N only depends on the length of I.*

Proof. To see this, choose $N \in \mathbb{Z}$ such that $\frac{2^{-N-1}}{3} \leq |I| < \frac{2^{-N}}{3}$. Denote

$$A_N = \{(k2^{-N}); k \in \mathbb{Z}\}, \quad A'_N = \{((k+\frac{1}{3})2^{-N}, (k+\frac{2}{3})2^{-N}); k \in \mathbb{Z}\}.$$

Note that for any two points $a, b \in A_N \cup A'_N$, we have $|a - b| \geq \frac{1}{3}2^{-N} > |I|$. Thus there is no more than one element of $A_N \cup A'_N$ in I. Then $I \cap A_N = \phi$ or $I \cap A'_N = \phi$. Therefore, I must be contained in some $D_N^{k_I}$ or $D'^{k_I}_N$. ∎

Remark. See [21] for a generalization of Proposition 3.1.

Remark. If an $\mathcal{M}_c$-atom defined in Chapter 2 admits its supporting interval as D_N^k (resp. D'^k_N) for some $k, N \in \mathbb{Z}$, we call it $\mathcal{M}_c$-$\mathcal{D}$-atom (resp. $\mathcal{M}_c$-$\mathcal{D}'$-atom). Proposition 3.1 implies that an $\mathcal{M}_c$-atom is either an $\mathcal{M}_c$-$\mathcal{D}$-atom or an $\mathcal{M}_c$-$\mathcal{D}'$-atom up to a fixed factor. Therefore the atomic Hardy space $\mathcal{H}_c^{1,at}(\mathbb{R}, \mathcal{M})$ defined in Chapter 2 can be characterized only by $\mathcal{M}_c$-$\mathcal{D}$-atoms and $\mathcal{M}_c$-$\mathcal{D}'$-atoms. A similar remark applies to the atomic row Hardy space $\mathcal{H}_r^{1,at}(\mathbb{R}, \mathcal{M})$. See Chapter 5 for more results of this type.

The proof of the following Proposition (as well as that of Theorem 3.3) illustrates well our approach to reduce problems on functions to those on martingales.

Put
$$f_h(t) = \frac{1}{h_1 + h_2} \int_{t-h_1}^{t+h_2} f(x)dx, \quad \forall h = (h_1, h_2) \in \mathbb{R}^+ \times \mathbb{R}^+.$$

PROPOSITION 3.2. *Let $(a_n)_{n \in \mathbb{Z}}$ be a positive sequence in $L^p(L^\infty(\mathbb{R}) \otimes \mathcal{M})$ and $h_n = (h_{n,1}, h_{n,2}) \in \mathbb{R}^+ \times \mathbb{R}^+, n \in \mathbb{Z}$.*
  (i) *If $1 \le p < \infty$,*

$$\left\| \sum_{n \in \mathbb{Z}} (a_n)_{h_n} \right\|_{L^p(L^\infty(\mathbb{R}) \otimes \mathcal{M})} \le c_p \left\| \sum_{n \in \mathbb{Z}} a_n \right\|_{L^p(L^\infty(\mathbb{R}) \otimes \mathcal{M})}. \tag{3.6}$$

  (ii) *If $1 < p \le \infty$,*

$$\left\| \sup_{n \in \mathbb{Z}} |(a_n)_{h_n}| \right\|_{L^p(L^\infty(\mathbb{R}) \otimes \mathcal{M})} \le c_p \left\| \sup_{n \in \mathbb{Z}} |a_n| \right\|_{L^p(L^\infty(\mathbb{R}) \otimes \mathcal{M})}. \tag{3.7}$$

Proof. From Proposition 3.1, $\forall n \in \mathbb{Z}$, for every $t \in \mathbb{R}$, there exist some $k_t, N_n \in \mathbb{Z}$ such that $(t - h_{n,1}, t + h_{n,2})$ is contained in $D_{N_n}^{k_t}$ or $D'^{k_t}_{N_n}$ and

$$|D_{N_n}^{k_t}| = |D'^{k_t}_{N_n}| \le 6(h_{n,1} + h_{n,2}).$$

Thus

$$(a_n)_{h_n} \le 6(E(a_n|\mathcal{D}_{N_n}) + E(a_n|\mathcal{D}'_{N_n})), \quad \forall n \in \mathbb{Z}, \tag{3.8}$$

where $E(\cdot|\mathcal{D}_{N_n})$ (resp. $E(\cdot|\mathcal{D}'_{N_n})$) denotes the conditional expectation with respect to $\mathcal{D}_{N_n}$ (resp. $\mathcal{D}'_{N_n}$). Then (3.6) follows from Theorem 0.1 of [**14**]. By (3.2) and (3.6),

$$\left\| \sup_{n \in \mathbb{Z}} |(a_n)_{h_n}| \right\|_{L^p(L^\infty(\mathbb{R}) \otimes \mathcal{M})}$$
$$= \sup \Big\{ \sum_{n \in \mathbb{Z}} \tau \int_\mathbb{R} \frac{1}{h_{n,1} + h_{n,2}} \int_{t-h_{n,1}}^{t+h_{n,2}} a_n(x) dx \, b_n(t) dt : \left\| \sum_{n \in \mathbb{Z}} b_n \right\|_{L^q(L^\infty(\mathbb{R}) \otimes \mathcal{M})} \le 1 \Big\}$$
$$= \sup \Big\{ \sum_{n \in \mathbb{Z}} \tau \int_\mathbb{R} \frac{1}{h_{n,1} + h_{n,2}} \int_{x-h_{n,2}}^{x+h_{n,1}} b_n(t) dt \, a_n(x) dx : \left\| \sum_{n \in \mathbb{Z}} b_n \right\|_{L^q(L^\infty(\mathbb{R}) \otimes \mathcal{M})} \le 1 \Big\}$$
$$\le \sup \Big\{ \sum_{n \in \mathbb{Z}} \tau \int_\mathbb{R} b_n(x) a_n(x) dx : \left\| \sum_{n \in \mathbb{Z}} b_n \right\|_{L^q(L^\infty(\mathbb{R}) \otimes \mathcal{M})} \le c_p \Big\}$$
$$\le c_p \left\| \sup_{n \in \mathbb{Z}} |a_n| \right\|_{L^p(L^\infty(\mathbb{R}) \otimes \mathcal{M})}$$

This is (3.7). ∎

The following is our noncommutative Hardy-Littlewood maximal inequality. Denote by $\mathcal{P}(\mathcal{M})$ the family of all projections of a von Neumann algebra $\mathcal{M}$.

THEOREM 3.3. (i) *Let $f \in L^1(L^\infty(\mathbb{R}) \otimes \mathcal{M})$ and $\lambda > 0$. Then there exists $e^\lambda \in \mathcal{P}(L^\infty(\mathbb{R}) \otimes \mathcal{M})$ such that*

$$\sup_{h \in \mathbb{R}^+ \times \mathbb{R}^+} \left\| e^\lambda f_h e^\lambda \right\|_{L^\infty(\mathbb{R}) \otimes \mathcal{M}} \le \lambda, \quad \left[\tau \otimes \int \right](1 - e^\lambda) < \frac{c_1 \|f\|_1}{\lambda}. \tag{3.9}$$

1. THE NONCOMMUTATIVE HARDY-LITTLEWOOD MAXIMAL INEQUALITY

(ii) Let $1 < p \leq \infty$ and $f \in L^p(L^\infty(\mathbb{R}) \otimes \mathcal{M})$. Then

$$\left\| \sup_{h \in \mathbb{R}^+ \times \mathbb{R}^+} |f_h| \right\|_{L^p(L^\infty(\mathbb{R}) \otimes \mathcal{M})} \leq c_p \|f\|_{L^p(L^\infty(\mathbb{R}) \otimes \mathcal{M})}. \tag{3.10}$$

Moreover, for every positive $f \in L^p(L^\infty(\mathbb{R}) \otimes \mathcal{M})$, there exists a positive $F \in L^p(L^\infty(\mathbb{R}) \otimes \mathcal{M})$ such that $f_h \leq F$ for all h and

$$\|F\|_{L^p(L^\infty(\mathbb{R}) \otimes \mathcal{M})} \leq c_p \|f\|_{L^p(L^\infty(\mathbb{R}) \otimes \mathcal{M})}. \tag{3.11}$$

Proof. By decomposing $f = f_1 - f_2 + i(f_3 - f_4)$ with positive f_k, we can assume f positive. To prove (i), for given $f, \lambda, (h_n)_{n \in \mathbb{Z}} \in \mathbb{R}^+ \times \mathbb{R}^+$, let $\mathcal{D}_{N_n}, \mathcal{D}'_{N_n}$ be as in the proof of Proposition 3.2. By the weak type (1,1) inequality of noncommutative martingales in [3] we have $\forall \lambda > 0$, $\exists e^\lambda, e'^\lambda \in \mathcal{P}(L^\infty(\mathbb{R}) \otimes \mathcal{M})$ such that

$$\sup_n \left\| e^\lambda E(f|\mathcal{D}_{N_n}) e^\lambda \right\|_{L^\infty(\mathbb{R}) \otimes \mathcal{M}} \leq \frac{\lambda}{12}, \quad \tau \otimes \int (1 - e^\lambda) < \frac{c\|f\|_1}{\lambda}$$

and

$$\sup_n \left\| e'^\lambda E(f|\mathcal{D}'_{N_n}) e'^\lambda \right\|_{L^\infty(\mathbb{R}) \otimes \mathcal{M}} \leq \frac{\lambda}{12}, \quad \tau \otimes \int (1 - e'^\lambda) < \frac{c\|f\|_1}{\lambda}$$

for every $f \in L^1(L^\infty(\mathbb{R}) \otimes \mathcal{M})$ and $(h_n)_{n \in \mathbb{Z}} \in \mathbb{R}^+ \times \mathbb{R}^+$. Let $\widetilde{e^\lambda} = e^\lambda \wedge e'^\lambda$, then

$$\tau \otimes \int (1 - \widetilde{e^\lambda}) < \frac{2c\|f\|_1}{\lambda}.$$

By Proposition 3.1, we have

$$\widetilde{e^\lambda} f_{h_n} \widetilde{e^\lambda} \leq 6(e^\lambda E(f|\mathcal{D}_{N_n}) e^\lambda + e'^\lambda E(f|\mathcal{D}'_{N_{h_n}}) e'^\lambda).$$

Therefore,

$$\sup_{h \in \mathbb{R}^+ \times \mathbb{R}^+} \left\| \widetilde{e^\lambda} f_h \widetilde{e^\lambda} \right\|_{L^\infty(\mathbb{R}) \otimes \mathcal{M}}$$
$$= \sup_{(h_n)_{n \in \mathbb{Z}}} \sup_n \left\| \widetilde{e^\lambda} f_{h_n} \widetilde{e^\lambda} \right\|_{L^\infty(\mathbb{R}) \otimes \mathcal{M}}$$
$$\leq 6 \sup_n \left\| e'^\lambda E(f|\mathcal{D}'_{N_n}) e'^\lambda \right\|_{L^\infty(\mathbb{R}) \otimes \mathcal{M}} + 6 \sup_n \left\| e^\lambda E(f|\mathcal{D}_{N_n}) e^\lambda \right\|_{L^\infty(\mathbb{R}) \otimes \mathcal{M}}$$
$$\leq \lambda.$$

This is (3.9). To prove (3.10), consider the two filtrations $\mathcal{D}, \mathcal{D}'$ introduced above. By Theorem 0.2 of [14], there exist two positive $F_1, F_2 \in L^p(L^\infty(\mathbb{R}) \otimes \mathcal{M})$ such that $\|F_1\|_{L^p}, \|F_2\|_{L^p} \leq c_p \|f\|_{L^p}$, and

$$E(f|\mathcal{D}_n) \leq F_1, \quad \text{and} \quad E(f|\mathcal{D}'_n) \leq F_2, \quad \forall n \in \mathbb{Z}. \tag{3.12}$$

Thus, similar to (3.8), we have (by Proposition 3.1), for every $h \in \mathbb{R}^+ \times \mathbb{R}^+$,

$$f_h \leq 6(F_1 + F_2) \tag{3.13}$$

Let $F = 6(F_1 + F_2)$, we proved (3.11). (3.10) follows immediately by decomposing $f = f_1 - f_2 + i(f_3 - f_4)$ with positive f_k. ∎

Using standard arguments and Theorem 3.3 we can easily obtain the noncommutative analogue of the classical non-tangential maximal inequality. Recall, as in Chapter 1, we also use f to denote its Poisson integral on the upper half plane.

THEOREM 3.4. *(i)* Let $f \in L^1(L^\infty(\mathbb{R}) \bar{\otimes} \mathcal{M})$. Then $\forall \lambda > 0, \exists\, e^\lambda \in \mathcal{P}(L^\infty(\mathbb{R}) \bar{\otimes} \mathcal{M})$, such that

(3.14) $\displaystyle\sup_{(t,y)\in\Gamma} \|e^\lambda f(x+t,y)e^\lambda\|_{L^\infty(\mathbb{R})\bar{\otimes}\mathcal{M}} \le \lambda,\quad \tau \otimes \int (1-e^\lambda) < \dfrac{c_1 \|f\|_1}{\lambda}, \forall \lambda > 0$

(ii) Let $f \in L^p(L^\infty(\mathbb{R}) \bar{\otimes} \mathcal{M}), 1 < p \le \infty$. Then

(3.15) $\left\|\displaystyle\sup_{(t,y)\in\Gamma} |f(x+t,y)|\right\|_p \le c_p \|f\|_p.$

Moreover, for every positive $f \in L^p(L^\infty(\mathbb{R}) \bar{\otimes} \mathcal{M})$, there exists a positive $F \in L^p(L^\infty(\mathbb{R}) \bar{\otimes} \mathcal{M})$ such that $f(\cdot + t, y) \le F$ for all $(t,y) \in \Gamma$ and

(3.16) $\|F\|_p \le c_p \|f\|_p.$

Proof. Notice that
$$P_y(x) = \frac{1}{\pi}\frac{y}{x^2+y^2} \le \frac{1}{\pi}\frac{1}{2^{2(k-1)}y + y},\quad \forall 2^{k-1}y \le |x|.$$

We have, for every positive f and any $(t,y) \in \Gamma$,

$$\begin{aligned}
& f(x+t,y) \\
&= \int_\mathbb{R} f(s) P_y(x+t-s) ds \\
&\le \frac{1}{\pi}\int_{|x+t-s|\le y} f(s) \frac{1}{y} ds + \frac{1}{\pi}\sum_{k=1}^\infty \int_{2^{k-1}y \le |x+t-s| \le 2^k y} f(s) \frac{1}{2^{2(k-1)}y + y} ds
\end{aligned}$$

(3.17) $\le \dfrac{1}{\pi}\displaystyle\sum_{k=0}^\infty \dfrac{8}{2^k}\dfrac{1}{2^{k+1}y}\int_{|x+t-s|\le 2^k y} f(s) ds.$

Considering $h_{k,y} = (2^k y - t, 2^k y + t) \in \mathbb{R}^+ \times \mathbb{R}^+$, we get (3.16) from (3.11). And by (3.10),

$$\left\|\sup_{(t,y)\in\Gamma} |f(x+t,y)|\right\|_p \le \frac{1}{\pi}\sum_{k=0}^\infty \frac{8}{2^k} \left\|\sup_{h_{k,y}} |f_{h_{k,y}}|\right\|_p$$
$$\le c_p \|f\|_p.$$

Decomposing $f = f_1 - f_2 + i(f_3 - f_4)$ with positive f_k, we get (3.15) for all $f \in L^p(L^\infty(\mathbb{R}) \bar{\otimes} \mathcal{M})$. We can prove (3.14) similarly. ∎

2. The noncommutative Lebesgue differentiation theorem and non-tangential limit of Poisson integrals

We end this chapter with the noncommutative Lebesgue differentiation theorem and non-tangential limit of Poisson integrals. These are consequences of Theorem 3.3 and Theorem 3.4. To this end, we first need to recall the noncommutative version of the almost everywhere convergence. Let $(f_\lambda)_{\lambda \in \Lambda}$ be a family of elements in $L^p(\mathcal{M}, \tau)$. We say $(f_\lambda)_{\lambda \in \Lambda}$ converges to f almost uniformly, abbreviated as $f_\lambda \overset{a.u}{\to} f$, if for every $\varepsilon > 0$, there exists $e_\varepsilon \in \mathcal{P}(\mathcal{M})$ such that $\tau(1 - e_\varepsilon) < \varepsilon$ and

$$\lim_{\lambda \to \lambda_0} \|e_\varepsilon (f_\lambda - f)\|_\infty = 0.$$

Moreover, we say $(f_\lambda)_{\lambda \in \lambda}$ converges to f bilaterally almost uniformly, abbreviated as $f_\lambda \overset{b.a.u}{\to} f$, if for every $\varepsilon > 0$, there exists $e_\varepsilon \in \mathcal{P}(\mathcal{M})$ such that $\tau(1 - e_\varepsilon) < \varepsilon$ and
$$\lim_{\lambda \to \lambda_0} \|e_\varepsilon (f_\lambda - f) e_\varepsilon\|_\infty = 0.$$

Obviously, $f_\lambda \overset{a.u}{\to} f$ implies $f_\lambda \overset{b.a.u}{\to} f$.

Recall that the map $x \mapsto x^p$ $(1 \le p \le 2)$ is convex on the positive cone $\mathcal{M}_+$ of $\mathcal{M}$ (see [**2**]). Thus, for $f \in L^p(L^\infty(\mathbb{R}) \otimes \mathcal{M})$ $(1 \le p \le 2)$, we get

(3.18) $$\int_A |f| dt \le (\int_A |f|^p dt)^{\frac{1}{p}}, \quad \forall A \subseteq \mathbb{R}, \ |A| = 1.$$

Note that for any $x, y \in \mathcal{M}_+$, $x \le y$ implies $x^q \le y^q, \forall 0 < q \le 1$. Using (3.18) successively, we get the following Lemma.

LEMMA 3.5. *For $f \in L^p(L^\infty(\mathbb{R}) \otimes \mathcal{M})$, $1 \le p < \infty$,*

(3.19) $$\int_A |f| dt \le (\int_A |f|^p dt)^{\frac{1}{p}}, \quad \forall A \subseteq \mathbb{R}, \ |A| = 1.$$

Recall that for any bounded linear operators a, b on a Hilbert space H, a positive and $\|b\| \le 1$, if T is an operator monotone function defined for positive operators (for example, $T(a) = a^{\frac{1}{p}}, p \ge 1$) then

(3.20) $$b^* T(a) b \le T(b^* a b).$$

This is the so-called Hansen's inequality (see [**9**]). In particular, we have

(3.21) $$b^* a b \le (b^* a^p b)^{\frac{1}{p}}.$$

THEOREM 3.6. *(i) Let $1 \le p < 2$. We have $f_h \overset{b.a.u}{\to} f$ as $h \to 0$ for any $f \in L^p(L^\infty(\mathbb{R}) \otimes \mathcal{M})$.*
(ii) Let $2 \le p < \infty$. We have $f_h \overset{a.u}{\to} f$ as $h \to 0$ for any $f \in L^p(L^\infty(\mathbb{R}) \otimes \mathcal{M})$.

Proof. (i) Without loss of generality, we can assume f selfadjoint. For any given $f \in L^p(L^\infty(\mathbb{R}) \otimes \mathcal{M})$ and $\varepsilon > 0$, choose $f^n = \sum_{k=1}^{N_n} \varphi_k x_k$, where $x_k \in S_\mathcal{M}^+$ and where $\varphi_k : \mathbb{R} \to \mathbb{C}$ are continuous functions with compact support, such that

(3.22) $$\||f - f^n|^p\|_1 = \|f - f^n\|_p^p < (\frac{1}{2^n})^p \frac{\varepsilon}{2^n}.$$

Choose $e_{1,n}^\varepsilon \in \mathcal{P}(L^\infty(\mathbb{R}) \otimes \mathcal{M})$ such that

$$\tau \otimes \int (1 - e_{1,n}^\varepsilon) < \frac{\varepsilon}{2^n} \quad \text{and} \quad \|e_{1,n}^\varepsilon |f^n - f|^p e_{1,n}^\varepsilon\|_{L^\infty(\mathbb{R}) \otimes \mathcal{M}} < (\frac{1}{2^n})^p.$$

Set $e_1^\varepsilon = \wedge_n e_{1,n}^\varepsilon$. We have $\tau \otimes \int (1 - e_1^\varepsilon) < \varepsilon$ and by (3.21),

$$\|e_1^\varepsilon (f^n - f) e_1^\varepsilon\|_{L^\infty(\mathbb{R}) \otimes \mathcal{M}} \le \|e_1^\varepsilon |f^n - f| e_1^\varepsilon\|_{L^\infty(\mathbb{R}) \otimes \mathcal{M}}$$
$$\le \|e_1^\varepsilon |f^n - f|^p e_1^\varepsilon\|_{L^\infty(\mathbb{R}) \otimes \mathcal{M}}^{\frac{1}{p}}$$
(3.23) $$< \frac{1}{2^n}, \ \forall n \ge 1.$$

On the other hand, by (3.9) and (3.22) we can find a sequence $(e_{2,n}^\varepsilon)_{n\geq 0} \subset \mathcal{P}(L^\infty(\mathbb{R})\otimes \mathcal{M})$ such that

$$\tau \otimes \int (1 - e_{2,n}^\varepsilon) < \frac{\varepsilon}{2^n}$$

(3.24) $\quad \left\|e_{2,n}^\varepsilon(|f^n - f|^p)_h e_{2,n}^\varepsilon\right\|_{L^\infty(\mathbb{R})\otimes \mathcal{M}} < (\frac{1}{2^n})^p, \quad \forall h \in \mathbb{R}^+ \times \mathbb{R}^+.$

Set $e_2^\varepsilon = \wedge_n e_{2,n}^\varepsilon$, we have $\tau \otimes \int (1 - e_2^\varepsilon) < \varepsilon$. By (3.19), (3.21) and (3.24)

$$\begin{aligned}
\|e_2^\varepsilon(f_h^n - f_h)e_2^\varepsilon\|_{L^\infty(\mathbb{R})\otimes \mathcal{M}} &\leq \left\|e_{2,n}^\varepsilon(|f^n - f|)_h e_{2,n}^\varepsilon\right\|_{L^\infty(\mathbb{R})\otimes \mathcal{M}} \\
&\leq \left\|e_{2,n}^\varepsilon(|f^n - f|^p)_h^{\frac{1}{p}} e_{2,n}^\varepsilon\right\| \\
&\leq (\left\|e_{2,n}^\varepsilon(|f^n - f|^p)_h e_{2,n}^\varepsilon\right\|_{L^\infty(\mathbb{R})\otimes \mathcal{M}})^{\frac{1}{p}}
\end{aligned}$$

(3.25) $\qquad\qquad\qquad < \dfrac{1}{2^n}, \quad \forall n \geq 0, h \in \mathbb{R}^+ \times \mathbb{R}^+.$

Recall that by the classical Lebesgue differentiation theorem,

$$\lim_{h\to 0} \|\varphi_h - \varphi\|_\infty = 0$$

if $\varphi : \mathbb{R} \to \mathbb{C}$ is continuous with compact support. Then by the choice of f_n we deduce that

$$\lim_{h\to 0} \|f_h^n - f^n\|_{L^\infty(\mathbb{R})\otimes \mathcal{M}} = 0, \forall n \geq 1.$$

Let $e^\varepsilon = e_1^\varepsilon \wedge e_2^\varepsilon$, then $\tau \otimes \int (1 - e^\varepsilon) < 2\varepsilon$. For any $n > 0$, choose $S_n > 0$ such that $\|f_h^n - f^n\|_\infty < \frac{1}{2^n}$ for any $h \in \mathbb{R}^+ \times \mathbb{R}^+$ such that $h_1 + h_2 < S_n$. Then, for any $h \in \mathbb{R}^+ \times \mathbb{R}^+$ such that $h_1 + h_2 < S_n$,

$$\begin{aligned}
\|e^\varepsilon(f_h - f)e^\varepsilon\|_\infty &\leq \|e^\varepsilon(f^n - f)e^\varepsilon\|_\infty + \|f_h^n - f^n\|_\infty + \|e^\varepsilon(f_h^n - f_h)e^\varepsilon\|_\infty \\
&\leq \|e_1^\varepsilon(f^n - f)e_1^\varepsilon\|_\infty + \|f_h^n - f^n\|_\infty + \|e_2^\varepsilon(f_h^n - f_h)e_2^\varepsilon\|_\infty \\
&\leq \frac{3}{2^n}.
\end{aligned}$$

Thus $\lim_{h\to 0} \|e^\varepsilon(f_h - f)e^\varepsilon\|_\infty \to 0$. This completes the proof of (i).

(ii) The proof of (i) works well for the part (ii) of the theorem with some minor changes. Let $(f^n)_{n\in\mathbb{N}}$ and $e_1^\varepsilon, e_2^\varepsilon, e^\varepsilon$ be as above. Since $p \geq 2$, instead of (3.23), (3.25), by (3.19) and (3.21) we have

(3.26) $\quad \|e_1^\varepsilon(f^n - f)\|_\infty = \left\|e_1^\varepsilon|f^n - f|^2 e_1^\varepsilon\right\|_\infty^{\frac{1}{2}} \leq \|e_1^\varepsilon|f^n - f|^p e_1^\varepsilon\|_\infty^{\frac{1}{p}} < \dfrac{1}{2^n}, \forall n \geq 1;$

and also

$$\begin{aligned}
\|e_2^\varepsilon(f_h^n - f_h)\|_\infty &= \left\|e_2^\varepsilon|f_h^n - f_h|^2 e_2^\varepsilon\right\|_\infty^{\frac{1}{2}} \\
&\leq (\left\|e_2^\varepsilon(|f^n - f|^2)_h e_2^\varepsilon\right\|_\infty)^{\frac{1}{2}}
\end{aligned}$$

(3.27) $\qquad\qquad\qquad \leq (\left\|e_2^\varepsilon(|f^n - f|^p)_h e_2^\varepsilon\right\|_\infty)^{\frac{1}{p}} < \dfrac{1}{2^n}, \quad \forall n \geq 1.$

Then we can conclude as in the proof of (i). ∎

THEOREM 3.7. (i) Let $1 \leq p < 2, f \in L^p(L^\infty(\mathbb{R}) \otimes \mathcal{M})$. We have $f(\cdot + u, y) \stackrel{b.a.u}{\to} f$ as $\Gamma \ni (u, y) \to 0$.

(ii) Let $2 \leq p < \infty, f \in L^p(L^\infty(\mathbb{R}) \otimes \mathcal{M})$. We have $f(\cdot + u, y) \stackrel{a.u}{\to} f$ as $\Gamma \ni (u, y) \to 0$.

Proof. We can assume $f \geq 0$ by decomposing f into four positive parts. Given $\varepsilon > 0$, let $f^n, e^\varepsilon_{i,n}, e^\varepsilon_i$ ($i = 1, 2$) be as in the proof of Theorem 3.6. We use the same notation f^n for the Poisson integral of f^n. It is easy to see that

$$\lim_{(u,y)\to 0.} \|f^n(\cdot + u, y) - f^n\|_\infty \to 0, \quad \forall n \geq 0, \quad (u,y) \in \Gamma$$

Let $e^\varepsilon = e^\varepsilon_1 \wedge e^\varepsilon_2$. For any $n > 0$, choose $Y_n > 0$ such that

$$\|f^n(\cdot + u, y) - f^n\|_\infty < \frac{1}{2^n}$$

for any $(u, y) \in \Gamma, |u| + y \leq Y_n$. To prove (i), from (3.23), (3.25) we have, for any $(u, y) \in \Gamma, |u| + y \leq Y_n$,

$$\|e^\varepsilon(f(\cdot + u, y) - f(\cdot))e^\varepsilon\|_\infty$$
$$\leq \|e^\varepsilon(f^n - f)e^\varepsilon\|_\infty + \|f^n(\cdot + u, y) - f^n\|_\infty$$
$$\quad + \left\|e^\varepsilon(\int_\mathbb{R} (f - f^n)(s) P_y(x + u - s) ds)e^\varepsilon\right\|_\infty$$
$$\leq \frac{1}{2^n} + \frac{1}{2^n} + \sum_{k=0}^\infty \left\|e^\varepsilon(\int_{|x+u-s|\leq 2^k y} |f - f^n| \frac{2}{2^{2(k-1)}y + y} ds)e^\varepsilon\right\|_\infty$$
$$\leq \frac{2}{2^n} + \sum_{k=0}^\infty \frac{8}{2^k} \left\|e^\varepsilon_2 (\frac{1}{2^k y} \int_{|x+u-s|\leq 2^k y} |f - f^n| ds)e^\varepsilon_2\right\|_\infty$$
$$\leq \frac{2}{2^n} + \sum_{k=0}^\infty \frac{8}{2^k} \|e^\varepsilon_2(|f - f^n|)_{h_{k,y}} e^\varepsilon_2\|_\infty$$
$$\leq \frac{2}{2^n} + \frac{8}{2^n},$$

where $h_{k,y} = (2^k y - t, 2^k y + t) \in \mathbb{R}^+ \times \mathbb{R}^+$. Thus

$$\lim_{(u,y)\to 0} \|e^\varepsilon(f(\cdot + ty, y) - f)e^\varepsilon\|_\infty = 0, \forall \varepsilon > 0,$$

and then $f(\cdot + u, y) \stackrel{b.a.u}{\to} f$ when $\Gamma \ni (u, y) \to 0$. This is (i). Using (3.26) and (3.27) instead of (3.23) and (3.25), we can prove (ii) similarly. ∎

Remark. When $p = \infty$, the corresponding convergence problems discussed in this section are still open.

CHAPTER 4

The Duality between $\mathcal{H}^p$ and $\mathrm{BMO}^q, 1 < p < 2$.

In this chapter, we describe the dual of $\mathcal{H}^p_c(\mathbb{R}, \mathcal{M})$, which is $\mathrm{BMO}^q_c(\mathbb{R}, \mathcal{M})$ (q being the conjugate index of p), the latter is the L^q-space analogue of BMO space already considered in Chapters 1 and 2. These $\mathrm{BMO}^q_c(\mathbb{R}, \mathcal{M})$ spaces not only are used to describe the dual of $\mathcal{H}^p_c(\mathbb{R}, \mathcal{M})$ but also play an important role for all results in the sequel. In particular, we will use it to prove the map Ψ introduced in Chapter 3 extends to a bounded map from $L^p(L^\infty(\mathbb{R}) \otimes \mathcal{M}, L^2_c(\widetilde{\Gamma}))$ to $\mathcal{H}^p_c(\mathbb{R}, \mathcal{M})$ for all $1 < p < \infty$. Consequently, $\mathcal{H}^p_c(\mathbb{R}, \mathcal{M})$ can be considered as a complemented subspace of $L^p(L^\infty(\mathbb{R}) \otimes \mathcal{M}, L^2_c(\widetilde{\Gamma}))$. For the most part, our results in Chapter 4 are extension to the function space setting of results proved for noncommutative martingales in [16].

1. Operator valued BMO^q ($q > 2$)

Let $\varphi \in L^q(\mathcal{M}, L^2_c(\mathbb{R}, \frac{dt}{1+t^2}))$. For $h \in \mathbb{R}^+ \times \mathbb{R}^+$, denote $I_{h,t} = (t - h_1, t + h_2]$. Let

$$\varphi^{\#}_h(t) = \frac{1}{h_1 + h_2} \int_{I_{h,t}} |\varphi(x) - \varphi_{I_{h,t}}|^2 dx$$

Set, for $2 < q \leq \infty$,

$$\|\varphi\|_{\mathrm{BMO}^q_c} = \left\| \sup_{h \in \mathbb{R}^+ \times \mathbb{R}^+} |\varphi^{\#}_h| \right\|^{\frac{1}{2}}_{L^{\frac{q}{2}}(L^\infty(\mathbb{R}) \otimes \mathcal{M})}$$

and

$$\|\varphi\|_{\mathrm{BMO}^q_r} = \|\varphi^*\|_{\mathrm{BMO}^q_c}.$$

It is easy to check that $\|\cdot\|_{\mathrm{BMO}^q_c}$ and $\|\cdot\|_{\mathrm{BMO}^q_r}$ are norms. Let $\mathrm{BMO}^q_c(\mathbb{R}, \mathcal{M})$ (resp. $\mathrm{BMO}^q_r(\mathbb{R}, \mathcal{M})$) be the space of all $\varphi \in L^q(\mathcal{M}, L^2_c(\mathbb{R}, \frac{dt}{1+t^2}))$ (resp. $L^q(\mathcal{M}, L^2_r)$) such that $\|\varphi\|_{\mathrm{BMO}^q_c} < \infty$ (resp. $\|\varphi\|_{\mathrm{BMO}^q_r} < \infty$). $\mathrm{BMO}^q_{cr}(\mathbb{R}, \mathcal{M})$ is defined as the intersection of these two spaces

$$\mathrm{BMO}^q_{cr}(\mathbb{R}, \mathcal{M}) = \mathrm{BMO}^q_c(\mathbb{R}, \mathcal{M}) \cap \mathrm{BMO}^q_r(\mathbb{R}, \mathcal{M})$$

equipped with the norm

$$\|\varphi\|_{\mathrm{BMO}^q_{cr}} = \max\{\|\varphi\|_{\mathrm{BMO}^q_c}, \|\varphi\|_{\mathrm{BMO}^q_r}\}.$$

If $q = \infty$, all these spaces coincide with those introduced in Chapter 2. And if $\mathcal{M} = \mathbb{C}$, all these spaces coincide with the classical BMO^q. As in the case of $\mathrm{BMO}(\mathbb{R}, \mathcal{M})$, we regard $\mathrm{BMO}^q_c(\mathbb{R}, \mathcal{M})$ (resp. $\mathrm{BMO}^q_r(\mathbb{R}, \mathcal{M})$, $\mathrm{BMO}^q_r(\mathbb{R}, \mathcal{M})$) as normed spaces modulo constants. The following is the analogue for $\mathrm{BMO}^q_c(\mathbb{R}, \mathcal{M})$

of Proposition 1.3. Recall that $I_t^n = (t - 2^{n-1}, t + 2^{n-1}]$ for $t \in \mathbb{R}$ and $n \in \mathbb{Z}$. Note that we have trivially

(4.1) $$\left\| \frac{1}{2^k} \int_{I_t^k} |\varphi(s) - \varphi_{I_t^k}|^2 ds \right\|_{L^{\frac{q}{2}}(L^\infty(\mathbb{R}) \otimes \mathcal{M})}^{\frac{1}{2}} \leq \|\varphi\|_{\text{BMO}_c^q}$$

PROPOSITION 4.1. *Let* $2 < q \leq \infty$. *Let* $\varphi \in \text{BMO}_c^q(\mathbb{R}, \mathcal{M})$. *Then*

$$\|\varphi\|_{L^q(\mathcal{M}, L_c^2(\mathbb{R}, \frac{dt}{1+t^2}))} \leq c \left(\|\varphi\|_{\text{BMO}_c^q} + \left\|\varphi_{I_0^1}\right\|_{L^q(\mathcal{M})} \right).$$

Moreover, $\text{BMO}_c^q(\mathbb{R}, \mathcal{M}), \text{BMO}_r^q(\mathbb{R}, \mathcal{M}), \text{BMO}_{cr}^q(\mathbb{R}, \mathcal{M})$ *are Banach spaces.*

Proof. The proof is similar to that of Proposition 1.3. By (1.12) we have

$$|\varphi_{I_t^n} - \varphi_{I_0^1}|^2 \leq n \left(\sum_{k=3}^n |\varphi_{I_t^k} - \varphi_{I_t^{k-1}}|^2 + |\varphi_{I_t^2} - \varphi_{I_0^1}|^2 \right)$$

$$\leq n \left(\sum_{k=3}^n \frac{1}{2^{k-1}} \int_{I_t^{k-1}} |\varphi(s) - \varphi_{I_t^k}|^2 ds + \frac{1}{2} \int_{I_0^1} |\varphi(s) - \varphi_{I_t^2}|^2 ds \right)$$

$$\leq n \left(\sum_{k=3}^n \frac{2}{2^k} \int_{I_t^k} |\varphi(s) - \varphi_{I_t^k}|^2 ds + \frac{2}{4} \int_{I_t^2} |\varphi(s) - \varphi_{I_t^2}|^2 ds \right)$$

(4.2) $$= 2n \sum_{k=2}^n \frac{1}{2^k} \int_{I_t^k} |\varphi(s) - \varphi_{I_t^k}|^2 ds, \quad \forall n > 1, t \in [-1, 1].$$

Thus by (4.1)

(4.3) $$\left\| |\varphi_{I_t^n} - \varphi_{I_0^1}|^2 \right\|_{L^{\frac{q}{2}}(L^\infty(\mathbb{R}) \otimes \mathcal{M})} \leq 2n^2 \|\varphi\|_{\text{BMO}_c^q}^2, \quad \forall n > 1, t \in [-1, 1].$$

To control φ's $L^q(\mathcal{M}, L_c^2(\mathbb{R}, \frac{dt}{1+t^2}))$ norm by its BMO_c^q norm, we write

$$\|\varphi\|_{L^q(\mathcal{M}, L_c^2(\mathbb{R}, \frac{dt}{1+t^2}))}^2$$

$$= \left\| \int_\mathbb{R} \frac{|\varphi(s)|^2}{1+s^2} ds \right\|_{L^{\frac{q}{2}}(\mathcal{M})}$$

$$= \left\| \chi_{[-\frac{1}{2}, \frac{1}{2}]}(t) \int_\mathbb{R} \frac{|\varphi(s)|^2}{1+s^2} ds \right\|_{L^{\frac{q}{2}}(L^\infty(\mathbb{R}) \otimes \mathcal{M})}$$

$$\leq \left\| \chi_{[-\frac{1}{2}, \frac{1}{2}]}(t) \left(\sum_{n=0}^\infty \int_{I_t^{n+1}/I_t^n} \frac{|\varphi(s)|^2}{1+s^2} ds + \int_{I_0^1} \frac{|\varphi(s)|^2}{1+s^2} ds \right) \right\|_{L^{\frac{q}{2}}(L^\infty(\mathbb{R}) \otimes \mathcal{M})}$$

$$\leq c \left(\left\| \chi_{[-\frac{1}{2}, \frac{1}{2}]}(t) \left(\sum_{n=2}^\infty \int_{I_t^n} \frac{|\varphi(s)|^2}{2^{2n}} ds + \int_{I_0^1} |\varphi(s)|^2 ds \right) \right\|_{L^{\frac{q}{2}}(L^\infty(\mathbb{R}) \otimes \mathcal{M})}$$

hence by (4.3)

$$\|\varphi\|^2_{L^q(\mathcal{M}, L^2_c(\mathbb{R}, \frac{dt}{1+t^2}))} \leq c(\left\|\sum_{n=2}^{\infty} \chi_{[-\frac{1}{2}, \frac{1}{2}]}(t) \int_{I^n_t} \frac{|\varphi(s) - \varphi_{I^n_t}|^2}{2^{2n}} ds\right\|_{L^{\frac{q}{2}}(L^{\infty}(\mathbb{R}) \otimes \mathcal{M})}$$
$$+ \left\|\sum_{n=1}^{\infty} \frac{|\varphi_{I^n_0}|^2}{2^n}\right\|_{L^{\frac{q}{2}}(\mathcal{M})} + \sum_{n=1}^{\infty} \frac{n^2 \|\varphi\|^2_{\mathrm{BMO}^q_c}}{2^n}$$
$$\leq c \sum_{n=1}^{\infty} \frac{(n^2+1) \|\varphi\|^2_{\mathrm{BMO}^q_c}}{2^n} + c \left\|\varphi_{I^1_0}\right\|^2_{L^q(\mathcal{M})}$$

(4.4)
$$< \infty.$$

Thus $\mathrm{BMO}^q_c(\mathbb{R}, \mathcal{M})$ is a Banach space. Passing to adjoints we get that $\mathrm{BMO}^q_r(\mathbb{R}, \mathcal{M})$ is a Banach spaces and then so is $\mathrm{BMO}^q_{cr}(\mathbb{R}, \mathcal{M})$. ∎

Put

$$\lambda^{n,\#}_\varphi(t) = \frac{1}{2^n} \iint_{T(I^n_t)} |\nabla \varphi|^2 y dx dy.$$

LEMMA 4.2. *Let* $\varphi \in \mathrm{BMO}^q_c(\mathbb{R}, \mathcal{M})$ $(2 < q < \infty)$. *Then* $\exists c > 0$ *such that*

$$\left\|\sup_{n \in \mathbb{Z}} |\lambda \varphi^{n,\#}|\right\|_{L^{\frac{q}{2}}(L^\infty(\mathbb{R}) \otimes \mathcal{M})} \leq c \|\varphi\|^2_{\mathrm{BMO}^q_c}.$$

Proof. The proof is similar to that of Lemma 1.4 but more complicated. For any $n \in \mathbb{Z}, t \in \mathbb{R}$, write $\varphi = \varphi^{n,t}_1 + \varphi^{n,t}_2 + \varphi^{n,t}_3$, where $\varphi^{n,t}_1 = (\varphi - \varphi_{I^{n+1}_t})\chi_{I^{n+1}_t}$, $\varphi^{n,t}_2 = (\varphi - \varphi_{I^{n+1}_t})\chi_{(I^{n+1}_t)^c}$, and $\varphi^{n,t}_3 = \varphi_{I^{n+1}_t}$. Set

$$\lambda^{n,\#}_i(t) = \frac{1}{2^n} \iint_{T(I^n_t)} |\nabla \varphi^{n,t}_i|^2 y dx dy, \quad i = 1, 2.$$

Thus

$$\left\|\sup_{n \in \mathbb{Z}} |\lambda \varphi^{n,\#}|\right\|_{L^{\frac{q}{2}}(L^\infty(\mathbb{R}) \otimes \mathcal{M})}$$
$$\leq 2 \left\|\sup_{n \in \mathbb{Z}} |\lambda^{n,\#}_1|\right\|_{L^{\frac{q}{2}}(L^\infty(\mathbb{R}) \otimes \mathcal{M})} + 2 \left\|\sup_{n \in \mathbb{Z}} |\lambda^{n,\#}_2|\right\|_{L^{\frac{q}{2}}(L^\infty(\mathbb{R}) \otimes \mathcal{M})}.$$

We treat $\lambda^{n,\#}_1$ first. Arguing as earlier for (1.19), by Green's theorem we have

$$\frac{1}{2^n} \iint_{T(I^n_t)} |\nabla \varphi^{n,t}_1|^2 y dx dy \leq \frac{1}{2^n} \int_{-\infty}^{+\infty} |\varphi^{n,t}_1|^2 ds.$$

Therefore,

$$\left\|\sup_{n\in\mathbb{Z}}|\frac{1}{2^n}\iint_{T(I_t^n)}|\nabla\varphi_1^{n,t}|^2 y dx dy|\right\|_{L^{\frac{q}{2}}(L^\infty(\mathbb{R})\overline{\otimes}\mathcal{M})}$$

$$\leq \left\|\sup_{n\in\mathbb{Z}}|\frac{1}{2^n}\int_{-\infty}^{+\infty}|\varphi_1^{n,t}|^2 ds|\right\|_{L^{\frac{q}{2}}(L^\infty(\mathbb{R})\overline{\otimes}\mathcal{M})}$$

$$= \left\|\sup_{n\in\mathbb{Z}}|\frac{1}{2^n}\int_{I_t^{n+1}}|\varphi - \varphi_{I_t^{n+1}}|^2 ds|\right\|_{L^{\frac{q}{2}}(L^\infty(\mathbb{R})\overline{\otimes}\mathcal{M})}$$

(4.5) $$\leq 2\|\varphi\|_{\mathrm{BMO}_c^q}^2$$

To deal with $\lambda_2^{n,\#}$, we note that

$$|\nabla P_y(x-s)|^2 \leq \frac{1}{4(x-s)^4} \leq \frac{c}{2^{4(n+k)}}, \quad \forall s \in I_t^{n+k+1}/I_t^{n+k}, \quad (x,y) \in T(I_t^n).$$

Let $A_k = I_t^{n+k+1}/I_t^{n+k}$. Then by (1.14), (1.17) and (4.2)

$$\frac{1}{2^n}\iint_{T(I_t^n)}|\nabla\varphi_2^{n,t}|^2 y dx dy$$

$$= \frac{1}{2^n}\iint_{T(I_t^n)}|\nabla\int_{-\infty}^{+\infty}P_y(x-s)\varphi_2^{n,t}(s)ds|^2 y dx dy$$

$$\leq \frac{1}{2^n}\iint_{T(I_t^n)}\left(\sum_{k=1}^\infty\int_{A_k}|\nabla P_y(x-s)|^2 2^{2k}ds \sum_{k=1}^\infty\int_{A_k}\frac{1}{2^{2k}}|\varphi_2^{n,t}(s)|^2 dsy\right)dxdy$$

$$\leq \frac{c}{2^n}\iint_{T(I_t^n)}\frac{1}{2^{3n}}\sum_{k=1}^\infty\int_{A_k}\frac{1}{2^{2k}}|\varphi - \varphi_{I_t^{n+1}}|^2 dsydxdy$$

$$\leq \frac{c}{2^n}\sum_{k=1}^\infty\int_{A_k}\frac{2}{2^{2k}}(|\varphi - \varphi_{I_t^{n+k+1}}|^2 + |\varphi_{I_t^{n+k+1}} - \varphi_{I_t^{n+1}}|^2)ds$$

$$\leq c\sum_{k=1}^\infty\frac{1}{2^{2k+n}}\int_{A_k}|\varphi - \varphi_{I_t^{n+k+1}}|^2 ds + \sum_{k=1}^\infty\frac{c}{2^k}\sum_{i=1}^k\frac{2k}{2^{n+i}}\int_{I_t^{n+i}}|\varphi(u) - \varphi_{I_t^{n+i}}|^2 du$$

$$\leq cX_n + cY_n$$

where

$$X_n = \sum_{k=1}^\infty\frac{1}{2^{2k+n}}\int_{A_k}|\varphi - \varphi_{I_t^{n+k+1}}|^2 ds,$$

$$Y_n = \sum_{k=1}^\infty\frac{k}{2^k}\sum_{i=1}^k\frac{1}{2^{n+i}}\int_{I_t^{n+i}}|\varphi(s) - \varphi_{I_t^{n+i}}|^2 ds.$$

X_n, Y_n are estimated as follows. For X_n we have

$$\left\| \sup_{n \in \mathbb{Z}} |X_n| \right\|_{L^{\frac{q}{2}}(L^\infty(\mathbb{R}) \otimes \mathcal{M})}$$

$$= \left\| \sup_{n \in \mathbb{Z}} | \sum_{k=1}^{\infty} \frac{1}{2^k} \frac{1}{2^{n+k}} \int_{A_k} |\varphi - \varphi_{I_t^{n+k+1}}|^2 ds | \right\|_{L^{\frac{q}{2}}(L^\infty(\mathbb{R}) \otimes \mathcal{M})}$$

$$\leq \sum_{k=1}^{\infty} \frac{1}{2^k} \left\| \sup_{n \in \mathbb{Z}} | \frac{1}{2^{n+k}} \int_{I_t^{n+k+1}} |\varphi - \varphi_{I_t^{n+k+1}}|^2 ds | \right\|_{L^{\frac{q}{2}}(L^\infty(\mathbb{R}) \otimes \mathcal{M})}$$

$$\leq 2 \|\varphi\|^2_{\mathrm{BMO}_c^q}.$$

On the other hand,

$$\left\| \sup_{n \in \mathbb{Z}} |Y_n| \right\|_{L^{\frac{q}{2}}(L^\infty(\mathbb{R}) \otimes \mathcal{M})}$$

$$\leq \sum_{k=1}^{\infty} \frac{k}{2^k} \sum_{i=1}^{k} \left\| \sup_{n \in \mathbb{Z}} | \frac{1}{2^{n+i}} \int_{I_t^{n+i}} |\varphi(s) - \varphi_{I_t^{n+i}}|^2 ds | \right\|_{L^{\frac{q}{2}}(L^\infty(\mathbb{R}) \otimes \mathcal{M})}$$

$$\leq \sum_{k=1}^{\infty} \frac{k^2}{2^k} \|\varphi\|^2_{\mathrm{BMO}_c^q}$$

$$= 6 \|\varphi\|^2_{\mathrm{BMO}_c^q}.$$

Combining the preceding inequalities we get

$$\left\| \sup_{n \in \mathbb{Z}} |\lambda^{n,\#}_{\varphi_2}| \right\|_{L^{\frac{q}{2}}(L^\infty(\mathbb{R}) \otimes \mathcal{M})} \leq c \|\varphi\|^2_{\mathrm{BMO}_c^q},$$

which, together with (4.5), yields

$$\left\| \sup_{n \in \mathbb{Z}} |\lambda \varphi^{n,\#}| \right\|_{L^{\frac{q}{2}}(L^\infty(\mathbb{R}) \otimes \mathcal{M})} \leq c \|\varphi\|^2_{\mathrm{BMO}_c^q}. \blacksquare$$

Set

$$\varphi_n^\#(t) = \frac{1}{2^n} \int_{I_t^n} |\varphi(x) - \varphi_{I_t^n}|^2 dx$$

Notice that for every $h \in \mathbb{R}^+ \times \mathbb{R}^+$ there exists $n \in \mathbb{Z}$ such that $(t - h_1, t + h_2) \in I_t^n$ for every $t \in \mathbb{R}$ and $2^n \leq 4(h_1 + h_2)$, we have

(4.6) $$\frac{1}{4} \|\varphi\|_{\mathrm{BMO}_c^q} \leq \left\| \sup_n \varphi_n^\# \right\|^{\frac{1}{2}}_{L^{\frac{q}{2}}(L^\infty(\mathbb{R}) \otimes \mathcal{M})} \leq \|\varphi\|_{\mathrm{BMO}_c^q}.$$

LEMMA 4.3. *The operator Ψ defined in Chapter 2 extends to a bounded map from $L^q(L^\infty(\mathbb{R}) \otimes \mathcal{M}, L_c^2(\widetilde{\Gamma}))$ $(2 < q < \infty)$ into $\mathrm{BMO}_c^q(\mathbb{R}, \mathcal{M})$ and there exists $c_q > 0$ such that*

(4.7) $$\|\Psi(h)\|_{\mathrm{BMO}_c^q} \leq c_q \|h\|_{L^q(L^\infty(\mathbb{R}) \otimes \mathcal{M}, L_c^2)}.$$

Proof. The pattern of this proof is similar to that of Lemma 2.2. One new thing we need is the noncommutative Hardy-Littlewood maximal inequality proved in the previous chapter.

Let $\mathcal{S}$ be the family of functions introduced in the proof of Lemma 2.2. Since $\mathcal{S}$ is dense in $L^q(L^\infty(\mathbb{R}) \bar{\otimes} \mathcal{M}, L^2_c(\widetilde{\Gamma}))$, we need only to prove (4.7) for all $h \in \mathcal{S}$. Fix $h \in \mathcal{S}$ and set $\varphi = \Psi(h)$. Then $\varphi \in L^q(\mathcal{M}, L^2_c(\mathbb{R}, \frac{ds}{1+s^2}))$. Let $u \in \mathbb{R}$ and $n \in \mathbb{Z}$. Set

$$h_1^u(x,y,t) = h(x,y,t)\chi_{I_u^{n+1}}(t),$$
$$h_2^u(x,y,t) = h(x,y,t)\chi_{(I_u^{n+1})^c}(t)$$

and

$$B_{I_u^n} = \int_{-\infty}^{+\infty} \iint_\Gamma Q_{I_u^n} h_2^u \, dy\, dx\, dt,$$

where

$$Q_{I_u^n}(x,y,t) = \frac{1}{2^n} \int_{I_u^n} Q_y(x+t-s)\, ds$$

(recall that $Q_y(x)$ is defined by (2.2) as the gradient of the Poisson kernel). Then

$$\begin{aligned}
\varphi_n^\#(u) &\leq \frac{4}{2^n} \int_{I_u^n} |\varphi(s) - B_{I_t^n}|^2 ds \\
&\leq \frac{8}{2^n} \int_{I_u^n} |\int_{(I_u^{n+1})^c} \iint_\Gamma (Q_y(x+t-s) - Q_{I_u^n})h \, dx\, dy\, dt|^2 ds \\
&\quad + \frac{8}{2^n} \int_{I_u^n} |\int_{-\infty}^{+\infty} \iint_\Gamma Q_y(x+t-s) h_1^u \, dx\, dy\, dt|^2 ds \\
&= 8A_n + \frac{8}{2^n} \int_{I_u^n} |\int_{I_u^{n+1}} \iint_\Gamma Q_y(x+t-s) h \, dx\, dy\, dt|^2 ds
\end{aligned}$$

Recall that, as noted earlier in (2.5),

$$\iint_\Gamma |Q_y(x+t-s) - Q_{I_u^n}|^2 dx\, dy \leq c 2^{2n}(t-u)^{-4}$$

for $t \in (I_u^{n+1})^c$ and $s \in I_u^n$. By (1.14), we have

$$\begin{aligned}
A_n &= \frac{1}{2^n} \int_{I_u^n} |\int_{(I_u^{n+1})^c} \iint_\Gamma (Q_y(x+t-s) - Q_{I_u^n}) h \, dx\, dy\, dt|^2 ds \\
&\leq \int_{(I_u^{n+1})^c} c 2^{2n}(t-u)^{-2} dt \int_{(I_u^{n+1})^c} (t-u)^{-2} \iint_\Gamma |h|^2 dx\, dy\, dt \\
&= c 2^n \int_{(I_u^{n+1})^c} (t-u)^{-2} \iint_\Gamma |h|^2 dx\, dy\, dt
\end{aligned}$$

Then, for any positive $(a_n)_{n\in\mathbb{Z}}$ such that $\left\|\sum_{k\in\mathbb{Z}} a_n\right\|_{L^{(\frac{q}{2})'}(L^\infty(\mathbb{R})\overline{\otimes}\mathcal{M})} \leq 1$,

$$\tau \sum_{n\in\mathbb{Z}} \int_{-\infty}^{+\infty} \varphi_n^{\#}(u) a_n(u) du$$
$$\leq \sum_{n\in\mathbb{Z}} \tau \int_{-\infty}^{+\infty} c2^n \int_{(I_u^{n+1})^c} (t-u)^{-2} \iint_\Gamma |h|^2 dx dy dt\, a_n(u) du$$
$$+ \sum_{n\in\mathbb{Z}} \tau \int_{-\infty}^{+\infty} \frac{8}{2^n} \int_{I_u^n} |\int_{I_u^{n+1}} \iint_\Gamma Q_y(x+t-s) h\, dx dy dt|^2 ds\, a_n(u) du$$
$$= A + B$$

By the noncommutative Hölder inequality,

$$A = \sum_{n\in\mathbb{Z}} \tau \int_{-\infty}^{+\infty} c2^n \int_{(I_t^{n+1})^c} (t-u)^{-2} a_n(u) du \iint_\Gamma |h|^2 dx dy dt$$
$$\leq \left\|\iint_\Gamma |h|^2 dx dy\right\|_{L^{\frac{q}{2}}(L^\infty(\mathbb{R})\overline{\otimes}\mathcal{M})} \left\|\sum_{n\in\mathbb{Z}} c2^n \int_{(I_t^n)^c} \frac{a_n(u)}{(t-u)^2} du\right\|_{L^{(\frac{q}{2})'}(L^\infty(\mathbb{R})\overline{\otimes}\mathcal{M})}$$
$$\leq \|h\|^2_{L^q(L^\infty(\mathbb{R})\overline{\otimes}\mathcal{M}, L_c^2(\widetilde{\Gamma}))} \left\|\sum_{n\in\mathbb{Z}} \sum_{k=n}^{+\infty} 2^n \int_{I_t^{k+1}} \frac{1}{2^{2k}} a_n(u) du\right\|_{L^{(\frac{q}{2})'}(L^\infty(\mathbb{R})\overline{\otimes}\mathcal{M})}.$$

Let us estimate the second factor in the last term. By (3.6),

$$\left\|\sum_{n\in\mathbb{Z}} \sum_{k=n+1}^{+\infty} 2^n \int_{I_t^{k+1}} \frac{1}{2^{2k}} a_n(u) du\right\|_{L^{(\frac{q}{2})'}(L^\infty(\mathbb{R})\overline{\otimes}\mathcal{M})}$$
$$= \left\|\sum_{k\in\mathbb{Z}} \frac{1}{2^k} \int_{I_t^{k+1}} \sum_{n=-\infty}^{k-1} \frac{2^n}{2^k} a_n(u) du\right\|_{L^{(\frac{q}{2})'}(L^\infty(\mathbb{R})\overline{\otimes}\mathcal{M})}$$
$$\leq c_q \left\|\sum_{k\in\mathbb{Z}} \sum_{n=-\infty}^{k-1} \frac{2^n}{2^k} a_n\right\|_{L^{(\frac{q}{2})'}(L^\infty(\mathbb{R})\overline{\otimes}\mathcal{M})}$$
$$\leq c_q \left\|\sum_{n\in\mathbb{Z}} a_n\right\|_{L^{(\frac{q}{2})'}(L^\infty(\mathbb{R})\overline{\otimes}\mathcal{M})} \leq c_q.$$

Thus

$$A \leq c_q \|h\|^2_{L^q(L^\infty(\mathbb{R})\overline{\otimes}\mathcal{M}, L_c^2)}.$$

For the term B, by (3.6), (1.10) and the Cauchy-Schwarz inequality,

$$B \leq \sum_{n\in\mathbb{Z}} \int_{\mathbb{R}} \frac{8}{2^n} \tau \int_{\mathbb{R}} |\int_{I_u^{n+1}} \iint_{\Gamma} Q_y(x+t-s)h dx dy dt|^2 ds a_n(u) du$$

$$= \sum_{n\in\mathbb{Z}} \int_{\mathbb{R}} \frac{8}{2^n} \sup_{\tau \int |f|^2 = 1} (\tau \int_{\mathbb{R}} \int_{I_u^{n+1}} \iint_{\Gamma} Q_y(x+t-s) h a_n^{\frac{1}{2}}(u) dx dy dt f(s) ds)^2 du$$

$$= \sum_{n\in\mathbb{Z}} \int_{\mathbb{R}} \frac{8}{2^n} \sup_{\tau \int |f|^2 = 1} (\tau \int_{I_u^{n+1}} \iint_{\Gamma} h a_n^{\frac{1}{2}}(u) \nabla f(t+x,y) dx dy dt)^2 du$$

$$\leq \sum_{n\in\mathbb{Z}} \int_{\mathbb{R}} \frac{8}{2^n} \tau \int_{I_u^{n+1}} \iint_{\Gamma} |h|^2 a_n(u) dx dy dt du$$

$$= \sum_{n\in\mathbb{Z}} \tau \int_{\mathbb{R}} \iint_{\Gamma} |h|^2 dx dy \frac{8}{2^n} \int_{I_t^{n+1}} a_n(u) du dt$$

$$\leq \| \iint_{\Gamma} |h|^2 dx dy \|_{L^{\frac{q}{2}}(L^\infty(\mathbb{R}) \otimes \mathcal{M})} \left\| \sum_{n\in\mathbb{Z}} \frac{16}{2^n} \int_{I_t^n} a_n(u) du \right\|_{L^{(\frac{q}{2})'}(L^\infty(\mathbb{R}) \otimes \mathcal{M})}$$

$$\leq c_q \|h\|^2_{L^q(L^\infty(\mathbb{R}) \otimes \mathcal{M}, L_c^2)}.$$

Thus

$$\left\| \sup_n |\varphi_n^{\#}| \right\|_{L^{\frac{q}{2}}(L^\infty(\mathbb{R}) \otimes \mathcal{M})} \leq c_q \|h\|^2_{L^q(L^\infty(\mathbb{R}) \otimes \mathcal{M}, L_c^2)}$$

and then

$$\|\Psi(h)\|_{\mathrm{BMO}_c^q} \leq c_q \|h\|_{L^q(L^\infty(\mathbb{R}) \otimes \mathcal{M}, L_c^2)}. \blacksquare$$

Remark. It seems difficult to define noncommutative BMO^q for $q < 2$.

2. The duality theorem of $\mathcal{H}^p$ and $\mathrm{BMO}^q (1 < p < 2)$

Denote by $\mathcal{H}_{c0}^p(\mathbb{R}, \mathcal{M})$ (resp. $\mathcal{H}_{r0}^p(\mathbb{R}, \mathcal{M})$) the functions f in $\mathcal{H}_c^p(\mathbb{R}, \mathcal{M})$ (resp. $\mathcal{H}_r^p(\mathbb{R}, \mathcal{M})$) such that $f \in L^p(\mathcal{M}, L_c^2(\mathbb{R}, (1+t^2)dt))$ (resp. $L^p(\mathcal{M}, L_r^2(\mathbb{R}, (1+t^2)dt))$) and $\int f dt = 0$. Set

$$\mathcal{H}_{cr0}^p(\mathbb{R}, \mathcal{M}) = \mathcal{H}_{c0}^p(\mathbb{R}, \mathcal{M}) + \mathcal{H}_{r0}^p(\mathbb{R}, \mathcal{M}).$$

It is easy to see that $\mathcal{H}_{c0}^p(\mathbb{R}, \mathcal{M})$ (resp. $\mathcal{H}_{r0}^p(\mathbb{R}, \mathcal{M})$, $\mathcal{H}_{cr0}^p(\mathbb{R}, \mathcal{M})$) is a dense subspace of $\mathcal{H}_c^p(\mathbb{R}, \mathcal{M})$ (resp. $\mathcal{H}_r^p(\mathbb{R}, \mathcal{H}_{cr0}^p(\mathbb{R}, \mathcal{M}))$. By Propositions 1.1 and 4.1, $\int_{-\infty}^{+\infty} \varphi^* f dt$ exists as an element in $L^1(\mathcal{M})$ for any $\varphi \in \mathrm{BMO}_c^q(\mathbb{R}, \mathcal{M})$ and $f \in \mathcal{H}_{c0}^p(\mathbb{R}, \mathcal{M})$.

THEOREM 4.4. *Let $1 < p < 2$, $q = \frac{p}{p-1}$. Then*

(a) $(\mathcal{H}_c^p(\mathbb{R}, \mathcal{M}))^* = \mathrm{BMO}_c^q(\mathbb{R}, \mathcal{M})$ with equivalent norms. More precisely, every $\varphi \in \mathrm{BMO}_c^q(\mathcal{M})$ defines a continuous linear functional on $\mathcal{H}_c^p(\mathbb{R}, \mathcal{M})$ by

$$(4.8) \qquad l\varphi(f) = \tau \int_{-\infty}^{+\infty} \varphi^* f dt; \qquad \forall f \in \mathcal{H}_{c0}^p(\mathbb{R}, \mathcal{M})$$

Conversely every $l \in (\mathcal{H}_c^p(\mathbb{R}, \mathcal{M}))^*$ can be given as above by some $\varphi \in \mathrm{BMO}_c^q(\mathbb{R}, \mathcal{M})$ and there exist constants $c, c_q > 0$ such that

$$c_q \|\varphi\|_{\mathrm{BMO}_c^q} \leq \|l\varphi\|_{(\mathcal{H}_c^p)^*} \leq c \|\varphi\|_{\mathrm{BMO}_c^q}$$

Thus $(\mathcal{H}_c^p(\mathbb{R}, \mathcal{M}))^* = \mathrm{BMO}_c^q(\mathbb{R}, \mathcal{M})$ *with equivalent norms.*

(b) Similarly, $(\mathcal{H}_r^p(\mathbb{R}, \mathcal{M}))^* = \mathrm{BMO}_r^q(\mathbb{R}, \mathcal{M})$ with equivalent norms.

(c) $(\mathcal{H}_{cr}^p(\mathbb{R}, \mathcal{M}))^* = \mathrm{BMO}_{cr}^q(\mathbb{R}, \mathcal{M})$ with equivalent norms.

Proof. (i) Let $\varphi \in \mathrm{BMO}_c^q(\mathbb{R}, \mathcal{M})$ and $f \in \mathcal{H}_{c0}^p(\mathbb{R}, \mathcal{M})$. As in the proof of Theorem 2.4, we assume φ and f compactly supported. Let $G_c(f)$ and $\widetilde{S}_c(f)$ be as in the proof of Theorem 2.4. Similar to what we have explained there, $G_c(f)(x, y)$ can be assumed to be invertible in $\mathcal{M}$ for every $(x, y) \in \mathbb{R}_+^2$. By Green's theorem and the Cauchy-Schwarz inequality (see the corresponding part of the proof of Theorem 2.4 to see why Green's theorem works well),

$$\begin{aligned} |l\varphi(f)| &= 2|\tau \int_{-\infty}^{+\infty} \int_0^\infty \nabla \varphi^* \nabla f y dy dx| \\ &\leq 2(\tau \int_{-\infty}^{+\infty} \int_0^\infty G_c^{p-2}(f)(x,y) |\nabla f|^2(x,y) y dy dx)^{\frac{1}{2}} \\ &\quad \bullet (3\tau \int_{-\infty}^{+\infty} \int_0^\infty \widetilde{S}_c^{2-p}(f)(x, \tfrac{y}{4}) |\nabla \varphi|^2 y dy dx)^{\frac{1}{2}} \\ &= 2I \bullet II \end{aligned}$$

Noting that $G_c^{p-1}(f)(x, y) \leq G_c^{p-1}(f)(x, 0)$, we have

$$\begin{aligned} I^2 &= \tau \int_{-\infty}^{+\infty} \int_0^\infty -G_c^{p-2}(f)(x,y) \frac{\partial G_c^2(f)}{\partial y}(x,y) dy dx \\ &= \tau \int_{-\infty}^{+\infty} \int_0^\infty (-G_c^{p-2}(f)(x,y) \frac{\partial G_c(f)}{\partial y} G_c(f)(x,y) \\ &\quad - G_c^{p-1}(f) \frac{\partial G_c(f)}{\partial y}(x,y)) dy dx \\ &= 2\tau \int_{-\infty}^{+\infty} \int_0^\infty -G_c^{p-1}(f)(x,y) \frac{\partial G_c(f)}{\partial y} dy dx \\ &\leq 2\tau \int_{-\infty}^{+\infty} \int_0^\infty -G_c^{p-1}(f)(x,0) \frac{\partial G_c(f)}{\partial y}(x,y) dx dy \\ &\leq 2\tau \int_{-\infty}^{+\infty} G_c^p(f)(x,0) dx \\ &\leq 6\tau \int_{-\infty}^{+\infty} S_c^p(f)(x) dx \\ &= 6 \|f\|_{\mathcal{H}_c^p}^p \end{aligned}$$

Define

$$\delta^k(x) = \widetilde{S}_c^{2-p}(f)(x, 2^k) - \widetilde{S}_c^{2-p}(f)(x, 2^{k+1}), \quad \forall x \in \mathbb{R}.$$

Then $\delta^k \in L^{\frac{p}{2-p}}(L^\infty(\mathbb{R}) \otimes \mathcal{M})$ is positive. Note that $(\frac{q}{2})' = \frac{p}{p-2}$. Moreover,

$$\begin{aligned} \delta^k(x) &= \delta^k(x'), \forall (i-1)2^j < x, x' \leq i2^j \\ \sum_{k=-\infty}^\infty \delta^k(x) &= \widetilde{S}_c^{2-p}(f)(x,0) \end{aligned}$$

Arguing as earlier for Theorem 2.4, we have

$$\begin{aligned}
II^2 &= 3\tau \int_{-\infty}^{+\infty} \sum_{k=-\infty}^{\infty} \widetilde{S}_c^{2-p}(f)(x,2^k) \int_{2^{k+2}}^{2^{k+3}} |\nabla\varphi|^2 y\,dy\,dx \\
&= 3\tau \int_{-\infty}^{+\infty} \sum_{k=-\infty}^{\infty} \Big(\sum_{j=k}^{\infty} \delta^j(x)\Big) \int_{2^{k+2}}^{2^{k+3}} |\nabla\varphi|^2 y\,dy\,dx \\
&= 3\tau \int_{-\infty}^{+\infty} \sum_{j=-\infty}^{\infty} 2^j \delta^j(x) \frac{1}{2^j} \int_0^{2^{j+3}} |\nabla\varphi|^2 y\,dy\,dx \\
&\le 3\tau \int_{-\infty}^{+\infty} \sum_{j=-\infty}^{\infty} \int_{x-2^j}^{x+2^j} \delta^j(t) dt \frac{1}{2^j} \int_0^{2^{j+3}} |\nabla\varphi|^2 y\,dy\,dx \\
&= 24\tau \sum_{j=-\infty}^{\infty} \int_{-\infty}^{+\infty} \delta^j(t) \frac{1}{2^{j+3}} \int_{t-2^j}^{t+2^j} \int_0^{2^{j+3}} |\nabla\varphi|^2 y\,dy\,dx\,dt
\end{aligned}$$

hence by (3.2) and Lemma 4.2

$$\begin{aligned}
II^2 &\le 24 \Big\|\sum_{j=-\infty}^{\infty} \delta^j(t)\Big\|_{L^{(\frac{q}{2})'}} \Big\|\sup_j |\frac{1}{2^{j+3}} \int_{t-2^j}^{t+2^j} \int_0^{2^{j+3}} |\nabla\varphi|^2 y\,dy\,dx|\Big\|_{L^{\frac{q}{2}}} \\
&\le c\|f\|_{\mathcal{H}_c^p}^{2-p} \|\varphi\|_{\mathrm{BMO}_c^q}^2.
\end{aligned}$$

Combining the preceding estimates on I and II, we get

$$|l_\varphi(f)| \le c\|\varphi\|_{\mathrm{BMO}_c^q} \|f\|_{\mathcal{H}_c^p}.$$

Therefore, $l\varphi$ defines a continuous functional on $\mathcal{H}_c^p$ of norm smaller than $c\|\varphi\|_{\mathrm{BMO}_c^q}$.

(ii) Now suppose $l \in (\mathcal{H}_c^p)^*$. Then by the Hahn-Banach theorem l extends to a continuous functional on $L^p(L^\infty(\mathbb{R}) \otimes \mathcal{M}, L_c^2(\widetilde{\Gamma}))$ of the same norm. Thus by

$$(L^p(L^\infty(\mathbb{R}) \otimes \mathcal{M}, L_c^2(\widetilde{\Gamma})))^* = L^q(L^\infty(\mathbb{R}) \otimes \mathcal{M}, L_c^2(\widetilde{\Gamma}))$$

there exists $h \in L^q(L^\infty(\mathbb{R}) \otimes \mathcal{M}, L_c^2(\widetilde{\Gamma}))$ such that

$$\|h\|^2_{L^q(L^\infty(\mathbb{R})\otimes\mathcal{M},L_c^2(\widetilde{\Gamma}))} = \|\iint_\Gamma h^*(x,y,t)h(x,y,t)dydx\|_{L^{\frac{q}{2}}(L^\infty(\mathbb{R})\otimes\mathcal{M})} = \|l\|^2$$

and

$$\begin{aligned}
l(f) &= \tau \int_{-\infty}^{+\infty} \iint_\Gamma h^*(x,y,t) \nabla f(t+x,y) dy\,dx\,dt \\
&= \tau \int_{-\infty}^{+\infty} \Psi^*(h) f(s) ds.
\end{aligned}$$

Let

(4.9) $$\varphi = \Psi(h)$$

Then

$$l(f) = \tau \int_{-\infty}^{+\infty} \varphi^*(s) f(s) ds$$

and by Lemma 4.3 $\|\varphi\|_{\mathrm{BMO}_c^q} \leq c_q\|l\|$. This finishes the proof of the theorem concerning $\mathcal{H}_c^p$ and BMO_c^q. Passing to adjoints yields the part on $\mathcal{H}_r^p$ and BMO_r^q. Finally, the duality between $\mathcal{H}_{cr}^p$ and BMO_{cr}^q is obtained by the classical fact that the dual of a sum is the intersection of the duals. ∎

COROLLARY 4.5. $\varphi \in \mathrm{BMO}_c^q(\mathbb{R}, \mathcal{M})$ if and only if

$$\left\|\sup_{n\in\mathbb{Z}} |\lambda\varphi^{n,\#}|\right\|_{L^{\frac{q}{2}}(L^\infty(\mathbb{R})\otimes\mathcal{M})} < \infty$$

and there exist $c, c_q > 0$ such that

$$c_q \|\varphi\|_{\mathrm{BMO}_c^q}^2 \leq \left\|\sup_{n\in\mathbb{Z}} |\lambda\varphi^{n,\#}|\right\|_{L^{\frac{q}{2}}(L^\infty(\mathbb{R})\otimes\mathcal{M})} \leq c \|\varphi\|_{\mathrm{BMO}_c^q}^2.$$

Proof. From the proof of Theorem 4.4, if φ is such that

$$\left\|\sup_{n} |\lambda_\varphi^{n,\#}|\right\|_{L^{\frac{q}{2}}(L^\infty(\mathbb{R})\otimes\mathcal{M})} < \infty,$$

then φ defines a continuous linear functional on $\mathcal{H}_{c0}^p$ by $l_\varphi = \tau \int_{-\infty}^{+\infty} \varphi^* f dt$ and

$$\|l_\varphi\|_{(\mathcal{H}_c^p)^*} \leq c \left\|\sup_{n} |\lambda_\varphi^{n,\#}|\right\|_{L^{\frac{q}{2}}(L^\infty(\mathbb{R})\otimes\mathcal{M})}^{\frac{1}{2}}$$

and then by Theorem 4.4 again, there exists a function $\varphi' \in \mathrm{BMO}_c^q(\mathbb{R}, \mathcal{M})$ with

$$\|\varphi'\|_{\mathrm{BMO}_c^q}^2 \leq c_q \|l_\varphi\|_{(\mathcal{H}_c^p)^*}^2 \leq c_q \left\|\sup_{n} \lambda_\varphi^{n,\#}\right\|_{L^{\frac{q}{2}}(L^\infty(\mathbb{R})\otimes\mathcal{M})}$$

such that

$$\tau \int_{-\infty}^{+\infty} \varphi^* f dt = \tau \int_{-\infty}^{+\infty} \varphi'^* f dt.$$

Thus $\varphi \in \mathrm{BMO}_c^q(\mathbb{R}, \mathcal{M})$ and $\|\varphi\|_{\mathrm{BMO}_c^q}^2 \leq c_q \left\|\sup_n \lambda_\varphi^{n,\#}\right\|_{L^{\frac{q}{2}}(L^\infty(\mathbb{R})\otimes\mathcal{M})}$. Combining this with Lemma 4.2, we get the desired assertion. ∎

Now we are in a position to show that as in the classical case, the Lusin square function and the Littlewood-Paley g-function have equivalent L^p-norm in the noncommutative setting. The case $p = 1$ was already obtained in Chapter 2.

THEOREM 4.6. For $f \in \mathcal{H}_c^p(\mathbb{R}, \mathcal{M})(resp. \mathcal{H}_r^p(\mathbb{R}, \mathcal{M}))$, $1 \leq p < \infty$, we have

(4.10) $\qquad c_p^{-1} \|G_c(f)\|_p \leq \|S_c(f)\|_p \leq c_p \|G_c(f)\|_p$;

(4.11) $\qquad c_p^{-1} \|G_r(f)\|_p \leq \|S_r(f)\|_p \leq c_p \|G_r(f)\|_p$.

Proof. We need only to prove the second inequality of (4.10). The case of $p = 2$ is obvious. The case of $p = 1$ is Corollary 2.7 and the part of $1 < p < 2$ can be proved similarly by using the following inequality already obtained during the proof of Theorem 4.4

$$|\tau \int \varphi^* f dt| \leq c \|\varphi\|_{\mathrm{BMO}_c^q} \|G_c(f)\|_p^{\frac{p}{2}} \|S_c(f)\|_p^{1-\frac{p}{2}}.$$

For $p > 2$, let g be a positive element in $L^{(\frac{p}{2})'}(L^\infty(\mathbb{R}) \otimes \mathcal{M})$ with $\|g\|_{(\frac{p}{2})'} \leq 1$. By (3.2) and (3.10) we have

$$\left| \tau \int_\mathbb{R} \iint_\Gamma |\nabla f(x+t,y)|^2 dxdy g(t)dt \right|$$

$$= \left| \tau \iint_{\mathbb{R}^2_+} |\nabla f(x,y)|^2 y \frac{1}{y} \int_{x-y}^{x+y} g(t) dt dx dy \right|$$

$$\leq 4 \left| \tau \int_\mathbb{R} \sum_{n=-\infty}^{+\infty} \int_{2^{n-1}}^{2^n} |\nabla f(x,y)|^2 y dy \frac{1}{2^{n+1}} \int_{x-2^n}^{x+2^n} g(t) dt dx \right|$$

$$\leq 4 \left\| \int_{\mathbb{R}_+} |\nabla f(x,y)|^2 y dy \right\|_{L^{\frac{p}{2}}(L^\infty(\mathbb{R}) \otimes \mathcal{M})} \left\| \sup_n \left| \frac{1}{2^{n+1}} \int_{x-2^n}^{x+2^n} g(t) dt \right| \right\|_{L^{(\frac{p}{2})'}(L^\infty(\mathbb{R}) \otimes \mathcal{M})}$$

$$\leq c_p \|G_c(f)\|_p^2$$

Therefore, taking the supremum over all g as above, we obtain

$$\|S_c(f)\|_p^2 \leq c_p \|G_c(f)\|_p^2. \quad \blacksquare$$

3. The equivalence of $\mathcal{H}^q$ and $\mathrm{BMO}^q (q > 2)$

The following is the analogue for functions of a result for noncommutative martingales proved in [**16**].

THEOREM 4.7. $\mathcal{H}_c^p(\mathbb{R}, \mathcal{M}) = \mathrm{BMO}_c^p(\mathbb{R}, \mathcal{M})$ *with equivalent norms for* $2 < p < \infty$.

Proof. Note that for every $\varphi \in \mathcal{H}_c^p(\mathbb{R}, \mathcal{M})$ and every $g \in \mathcal{H}_c^{p'}(\mathbb{R}, \mathcal{M})$ ($p' = \frac{p}{p-1}$)

$$\left| \tau \int_{-\infty}^{+\infty} \iint_\Gamma \nabla g(x+t,y) \nabla \varphi^*(x+t,y) dx dy dt \right|$$

$$\leq \|\nabla g(x+t,y)\|_{L^{p'}(L^\infty(\mathbb{R}) \otimes \mathcal{M}, L_c^2(\widetilde{\Gamma}))} \|\nabla \varphi(x+t,y)\|_{L^p(L^\infty(\mathbb{R}) \otimes \mathcal{M}, L_c^2(\widetilde{\Gamma}))}$$

$$\leq \|g\|_{\mathcal{H}_c^{p'}} \|\varphi\|_{\mathcal{H}_c^p}.$$

Then by Theorem 4.4

$$(4.12) \qquad \|\varphi\|_{\mathrm{BMO}_c^p} \leq c_p \sup_{\|g\|_{\mathcal{H}_c^{p'}} \leq 1} |\tau \int g \varphi^* dt| \leq c_p \|\varphi\|_{\mathcal{H}_c^p}.$$

To prove the converse, we consider the following tent space T_c^p. Denote $\widetilde{\mathbb{R}_+^2} = (\mathbb{R}_+^2, \frac{dxdy}{y^2}) \times (\{1,2\}, \sigma)$ with $\sigma\{1\} = \sigma\{2\} = 1$. For $f \in L^p(\mathcal{M}, L_c^2(\widetilde{\mathbb{R}_+^2}))$, set

$$A_c(f)(t) = \left(\iint_{|x|<y} |f(x+t,y)|^2 dx \frac{dy}{y^2} \right)^{\frac{1}{2}}.$$

Define, for $1 < p < \infty$,

$$(4.13) \qquad T_c^p = \{ f \in L^p(\mathcal{M}, L_c^2(\widetilde{\mathbb{R}_+^2})), \|f\|_{T_c^p} = \|A_c(f)\|_{L^p(L^\infty(\mathbb{R}) \otimes \mathcal{M})} < \infty \}.$$

We will prove that, for $p > 2$ and $\varphi \in \mathrm{BMO}_c^p(\mathbb{R}, \mathcal{M})$, φ induces a linear functional on $T_c^{p'}$ defined by

$$l_\varphi(f) = \tau \iint_{\mathbb{R}_+^2} \nabla \varphi^*(x,y) y f(x,y) dx dy/y$$

and

(4.14) $$\|\varphi\|_{\mathcal{H}_c^p} \leq c_p \|l_\varphi\| \leq c_p \|\varphi\|_{\mathrm{BMO}_c^p}.$$

We first prove the second inequality of (4.14). Set

$$A_c(f)(t,y) = \Big(\iint_{s>y, |x|<s-y} |f(x+t,s)|^2 dx \frac{ds}{s^2} \Big)^{\frac{1}{2}}$$

$$\overline{A}_c(f)(t,y) = \Big(\iint_{s>y, |x|<\frac{s}{4}} |f(x+t,s)|^2 dx \frac{ds}{s^2} \Big)^{\frac{1}{2}}.$$

It is easy to see that

(4.15) $$\overline{A}_c^2(f)(t,y) \leq \overline{A}_c^2(f)(t,0) \leq A_c^2(f)(t),$$

(4.16) $$\overline{A}_c^2(f)(t+x,y) \leq A_c^2(f)(t, \frac{y}{2}), \quad \forall |x| < \frac{y}{4}, (t,y) \in \mathbb{R}_+^2.$$

For nice f and by approximation, we can assume $A_c(f)(t,y)$ is invertible for all $(t,y) \in \mathbb{R}_+^2$. Thus by the Cauchy-Schwarz inequality

$$\begin{aligned} l_\varphi(f) &= \tau \iint_{\mathbb{R}_+^2} f(t,y) \nabla \varphi^*(t,y) y dt \frac{dy}{y} \\ &\leq \Big(\tau \iint_{\mathbb{R}_+^2} A_c^{p'-2}(f)(t,\frac{y}{2}) |f|^2 y dt \frac{dy}{y^2} \Big)^{\frac{1}{2}} \Big(\tau \iint_{\mathbb{R}_+^2} A_c^{2-p'}(f)(t,\frac{y}{2}) |\nabla \varphi|^2 y dt dy \Big)^{\frac{1}{2}} \\ &= I \cdot II \end{aligned}$$

Similarly to the proof of Theorem 4.4, we have

$$II^2 \leq c \|\varphi\|_{\mathrm{BMO}_c^p}^2 \|f\|_{T_c^{p'}}^{2-p'}$$

3. THE EQUIVALENCE OF $\mathcal{H}^q$ AND $\mathrm{BMO}^q (q > 2)$

Concerning the factor I, by (4.16) we have (recall $p' - 2 < 0$)

$$\begin{aligned}
I^2 &\leq \tau \iint_{\mathbb{R}^2_+} 2 \int_{t-\frac{y}{4}}^{t+\frac{y}{4}} \overline{A}_c^{p'-2}(f)(x,y)dx |f(t,y)|^2 dt \frac{dy}{y^2} \\
&\leq 2\tau \iint_{\mathbb{R}^2_+} \overline{A}_c^{p'-2}(f)(x,y) \int_{x-\frac{y}{4}}^{x+\frac{y}{4}} |f(t,y)|^2 dt dx \frac{dy}{y^2} \\
&\leq -2\tau \iint_{\mathbb{R}^2_+} \overline{A}_c^{p'-2}(f)(x,y) \frac{\partial \overline{A}_c^2(f)}{\partial y}(x,y) dy dx \\
&= -4\tau \iint_{\mathbb{R}^2_+} \overline{A}_c^{p'-1}(f)(x,y) \frac{\partial \overline{A}_c(f)}{\partial y}(x,y) dy dx \\
&\leq -4\tau \int_{\mathbb{R}} \overline{A}_c^{p'-1}(f)(x,0) \int_{\mathbb{R}^+} \frac{\partial \overline{A}_c(f)}{\partial y}(x,y) dy dx \\
&\leq 4 \|f\|_{T_c^{p'}}^{p'}
\end{aligned}$$

Thus

(4.17) $$\|l_\varphi\| \leq c \|\varphi\|_{\mathrm{BMO}_c^p}.$$

Next we prove that $\|\varphi\|_{\mathcal{H}_c^p} \leq c_p \|l_\varphi\|$. Since we can regard $T_c^{p'}$ as a closed subspace of $L^{p'}(L^\infty(\mathbb{R}) \otimes \mathcal{M}, L_c^2(\widetilde{\mathbb{R}^2_+}))$ via the map $f(x,y) \to f(x,y)\chi_{\{|x-t|<y\}}$. l_φ extends to a linear functional on $L^{p'}(L^\infty(\mathbb{R}) \otimes \mathcal{M}, L_c^2(\widetilde{\mathbb{R}^2_+}))$ with the same norm. Then there exists $h \in L^p(L^\infty(\mathbb{R}) \otimes \mathcal{M}, L_c^2(\widetilde{\mathbb{R}^2_+}))$ such that $\|h\|_{L^p(L^\infty(\mathbb{R}) \otimes \mathcal{M}, L_c^2(\widetilde{\mathbb{R}^2_+}))} \leq \|l_\varphi\|$ and

$$\begin{aligned}
l_\varphi(f) &= \tau \int_{\mathbb{R}} \iint_{|x-t|<y} f(x,y) h^*(x,y,t) dx \frac{dy}{y^2} dt \\
&= \tau \iint_{\mathbb{R}^2_+} f(x,y) \int_{x-y}^{x+y} h^*(x,y,t) dt dx \frac{dy}{y^2}.
\end{aligned}$$

for every $f(x,y) \in T_c^{p'}$. Thus

(4.18) $$\nabla \varphi(x,y) y = \frac{1}{y} \int_{x-y}^{x+y} h(x,y,t) dt.$$

Then

$$\|\varphi\|_{\mathcal{H}_c^p}^2 = (\tau \int_{\mathbb{R}} (\iint_{\Gamma} |\frac{1}{y} \int_{x+s-y}^{x+s+y} h(x+s,y,t)dt|^2 dx \frac{dy}{y^2})^{\frac{p}{2}} ds)^{\frac{2}{p}}$$

$$\leq (\tau \int_{\mathbb{R}} (\iint_{\mathbb{R}_+^2} \frac{1}{y} \int_{s-2y}^{s+2y} |h(x,y,t)|^2 dt dx \frac{dy}{y^2})^{\frac{p}{2}} ds)^{\frac{2}{p}}$$

$$= \left\| \iint_{\mathbb{R}_+^2} \frac{1}{y} \int_{s-2y}^{s+2y} |h(x,y,t)|^2 dt dx \frac{dy}{y^2} \right\|_{L^{\frac{p}{2}}(L^\infty(\mathbb{R})\overline{\otimes}\mathcal{M})}$$

Notice that, for every positive a with $\|a\|_{L^{(\frac{p}{2})'}(L^\infty(\mathbb{R})\overline{\otimes}\mathcal{M})} \leq 1$, by (3.10) and (3.2) we have

$$\tau \int_{\mathbb{R}} \iint_{\mathbb{R}_+^2} \frac{1}{y} \int_{s-2y}^{s+2y} |h(x,y,t)|^2 dt dx \frac{dy}{y^2} a(s) ds$$

$$= \tau \int_{\mathbb{R}} \iint_{\mathbb{R}_+^2} |h(x,y,t)|^2 \frac{1}{y} \int_{t-2y}^{t+2y} a(s) ds dx \frac{dy}{y^2} dt$$

$$\leq 8\tau \int_{\mathbb{R}} \sum_{n=-\infty}^{+\infty} \int_{2^{n-2}}^{2^{n-1}} \int_{\mathbb{R}} |h(x,y,t)|^2 dx \frac{dy}{y^2} \frac{1}{2^{n+1}} \int_{t-2^n}^{t+2^n} a(s) ds dt$$

$$\leq 8 \left\| \iint_{\mathbb{R}_+^2} |h(x,y,t)|^2 dx \frac{dy}{y^2} \right\|_{L^{\frac{p}{2}}(L^\infty(\mathbb{R})\overline{\otimes}\mathcal{M})} \left\| \sup_n |\frac{1}{2^{n+1}} \int_{t-2^n}^{t+2^n} a(s) ds| \right\|_{L^{(\frac{p}{2})'}(L^\infty(\mathbb{R})\overline{\otimes}\mathcal{M})}$$

$$\leq c_p \|h\|_{L^p(L^\infty(\mathbb{R})\overline{\otimes}\mathcal{M}, L_c^2(\widetilde{\mathbb{R}_+^2}))}^2 \leq c_p \|l_\varphi\|^2$$

Therefore by taking the supremum over all a as above, we obtain

$$\|\varphi\|_{\mathcal{H}_c^p}^2 \leq c_p \|l_\varphi\|^2$$

Combining this with (4.17) we get

$$\|\varphi\|_{\mathcal{H}_c^p} \leq c_p \|\varphi\|_{\mathrm{BMO}_c^p}.$$

Then $\|\varphi\|_{\mathcal{H}_c^p} \simeq \|\varphi\|_{\mathrm{BMO}_c^p}$ for every $\varphi \in \mathcal{H}_c^p(\mathbb{R}, \mathcal{M})$.

To prove $\mathrm{BMO}_c^p(\mathbb{R}, \mathcal{M})$ and $\mathcal{H}_c^p(\mathbb{R}, \mathcal{M})$ are the same space, it remains to show that the family of $S_\mathcal{M}$-simple functions is dense in $\mathrm{BMO}_c^p(\mathbb{R}, \mathcal{M})$. From the proof of Theorem 4.4 we can see that for every $\varphi \in \mathrm{BMO}_c^p(\mathbb{R}, \mathcal{M})$, there exists a $h \in L^\infty(L^\infty(\mathbb{R})\overline{\otimes}\mathcal{M}, L_c^2)$ such that $\varphi = \Psi(h)$ and $\|\Psi(h)\|_{\mathrm{BMO}_c^p} \leq c\|h\|_{L^p(L^\infty(\mathbb{R})\overline{\otimes}\mathcal{M}, L_c^2)}$. Recall that the family of "nice" h's (i.e. $h(x,y,t) = \sum_{i=1}^n m_i f_i(t) \chi_{A_i}$ with $m_i \in S_\mathcal{M}, A_i \in \widetilde{\Gamma}, |A_i| < \infty$ and with scalar valued simple functions f_i) is dense in $L^p(L^\infty(\mathbb{R})\overline{\otimes}\mathcal{M}, L_c^2)$. Choose "nice" $h_n \to h$ in $L^p(L^\infty(\mathbb{R})\overline{\otimes}\mathcal{M}, L_c^2)$. Let $\varphi_n = \Psi(h_n)$. Then $\varphi_n \to \varphi$ in $\mathrm{BMO}_c^p(\mathbb{R}, \mathcal{M})$. Since the φ_n's are continuous functions with compact support, we can approximate them by simple functions in $\mathrm{BMO}_c^p(\mathbb{R}, \mathcal{M})$. This shows the density of simple functions in $\mathrm{BMO}_c^p(\mathbb{R}, \mathcal{M})$ and thus completes the proof of the theorem. ∎

Remark. By the same idea used in the proof above, we can get the analogue of the classical duality result for the tent spaces: $(T_c^p)^* = T_c^q$ $(1 < p < \infty)$ with equivalent norms, where T_c^p is defined as (4.13).

THEOREM 4.8. *(i)* Ψ *extends to a bounded map from* $L^\infty(L^\infty(\mathbb{R}) \bar{\otimes} \mathcal{M}, L_c^2(\widetilde{\Gamma}))$ *into* $\mathrm{BMO}_c(\mathbb{R}, \mathcal{M})$ *and*

(4.19) $$\|\Psi(h)\|_{\mathrm{BMO}_c} \leq c \|h\|_{L^\infty(L^\infty(\mathbb{R}) \bar{\otimes} \mathcal{M}, L_c^2)}$$

(ii) Ψ *extends to a bounded map from* $L^p(L^\infty(\mathbb{R}) \bar{\otimes} \mathcal{M}, L_c^2(\widetilde{\Gamma}))$ *into* $\mathcal{H}_c^p(\mathbb{R}, \mathcal{M})$ $(1 < p < \infty)$ *and*

(4.20) $$\|\Psi(h)\|_{\mathcal{H}_c^p} \leq c_p \|h\|_{L^p(L^\infty(\mathbb{R}) \bar{\otimes} \mathcal{M}, L_c^2)}.$$

(iii) The statements (i) and (ii) also hold with column spaces replaced by row spaces.

Proof. (4.19) is Lemma 2.2. The part of (4.20) concerning $p > 2$ follows from Lemma 4.3 and Theorem 4.7. For $1 < p < 2$, by the duality between $\mathcal{H}_c^p$ and BMO_c^q, and Theorem 4.7, we have

$$\begin{aligned}
\|\Psi(h)\|_{\mathcal{H}_c^p} &\leq c \sup_{\|f\|_{\mathrm{BMO}_c^q} \leq 1} \left| \tau \int_\mathbb{R} \Psi(h)(s) f^*(s) ds \right| \\
&\leq \sup_{\|f\|_{\mathcal{H}_c^q} \leq 1} \left| \tau \int_\mathbb{R} \int_\mathbb{R} \iint_\Gamma h(x,y,t) \nabla P_y(x+t-s) dx dy dt \, f^*(s) ds \right| \\
&= \sup_{\|f\|_{\mathcal{H}_c^q} \leq c} \left| \tau \int_\mathbb{R} \iint_\Gamma h(x,y,t) \nabla f^*(x+t,y) dx dy dt \right|
\end{aligned}$$

(4.21) $\qquad \leq c \|h\|_{L^p(L^\infty(\mathbb{R}) \bar{\otimes} \mathcal{M}, L_c^2)}.$

When $p = 2$, similarly but taking supremum over $\|f\|_{\mathcal{H}_c^2} \leq 1$ in the formula above, we have $\|\Psi(h)\|_{\mathcal{H}_c^2} \leq \|h\|_{L^2(L^\infty(\mathbb{R}) \bar{\otimes} \mathcal{M}, L_c^2)}$.

COROLLARY 4.9. $(\mathcal{H}_c^p(\mathbb{R}, \mathcal{M}))^* = \mathcal{H}_c^q(\mathbb{R}, \mathcal{M})$ *with equivalent norms for all* $1 < p < \infty$.

CHAPTER 5

Reduction of BMO to dyadic BMO

Our approach in Chapter 3 towards the maximal inequality is to reduce it to the corresponding maximal inequality for dyadic martingales. In this chapter, we pursue this idea. We will see that BMO spaces can be characterized as intersections of dyadic BMO. This result has many consequences. It will be used in the next chapter for interpolation too.

1. BMO is the intersection of two dyadic BMO

Consider an increasing family of σ-algebras $\mathcal{F} = \{\mathcal{F}_n\}_{n \in \mathbb{Z}}$ on $\mathbb{R}$. Assume that each $\mathcal{F}_n$ is generated by a sequence of atoms $\{F_n^k\}_{k \in \mathbb{Z}}$. We are going to introduce the BMO^q spaces for martingales with respect to $\mathcal{F} = \{\mathcal{F}_n\}_{n \in \mathbb{Z}}$. Let $2 < q \le \infty$ and $\varphi \in L^q(\mathcal{M}, L_c^2(\mathbb{R}, \frac{dt}{1+t^2}))$. Define

$$\varphi_{\mathcal{F}_n}^{\#}(t) = \frac{1}{|F_n^k|} \int_{F_n^k \ni t} |\varphi(x) - \varphi_{F_n^k}|^2 dx$$

For $\varphi \in L^q(\mathcal{M}, L_c^2(\mathbb{R}, \frac{dt}{1+t^2}))$ (resp. $L^q(\mathcal{M}, L_r^2(\mathbb{R}, \frac{dt}{1+t^2}))$), let

$$\|\varphi\|_{\text{BMO}_c^{q,\mathcal{F}}} = \left\| \sup_n |\varphi_{\mathcal{F}_n}^{\#}| \right\|_{\frac{q}{2}}^{\frac{1}{2}} \quad \text{and} \quad \|\varphi\|_{\text{BMO}_r^{q,\mathcal{F}}} = \|\varphi^*\|_{\text{BMO}_c^{q,\mathcal{F}}}.$$

And set

$$\text{BMO}_c^{q,\mathcal{F}}(L^\infty(\mathbb{R}) \otimes \mathcal{M}) = \{\varphi \in L^q(\mathcal{M}, L_c^2(\mathbb{R}, \frac{dt}{1+t^2})), \|\varphi\|_{\text{BMO}_c^{q,\mathcal{F}}} < \infty\},$$

$$\text{BMO}_r^{q,\mathcal{F}}(L^\infty(\mathbb{R}) \otimes \mathcal{M}) = \{\varphi \in L^q(\mathcal{M}, L_r^2(\mathbb{R}, \frac{dt}{1+t^2})), \|\varphi\|_{\text{BMO}_r^{q,\mathcal{F}}} < \infty\}.$$

Define $\text{BMO}_{cr}^{q,\mathcal{F}}$ to be the intersection of $\text{BMO}_c^{q,\mathcal{F}}$ and $\text{BMO}_r^{q,\mathcal{F}}$ with the intersection norm $\max\{\|\varphi\|_{\text{BMO}_c^{q,\mathcal{F}}}, \|\varphi\|_{\text{BMO}_r^{q,\mathcal{F}}}\}$. These BMO^q spaces were already studied in [16] for general noncommutative martingales.

In the following, we will consider the dyadic BMO spaces $\text{BMO}_c^{q,\mathcal{D}}(L^\infty(\mathbb{R}) \otimes \mathcal{M})$, $\text{BMO}_c^{q,\mathcal{D}'}(L^\infty(\mathbb{R}) \otimes \mathcal{M})$, $\text{BMO}_r^{q,\mathcal{D}}(L^\infty(\mathbb{R}) \otimes \mathcal{M})$, $\text{BMO}_r^{q,\mathcal{D}'}(L^\infty(\mathbb{R}) \otimes \mathcal{M})$ etc. with respect to the families $\mathcal{D}, \mathcal{D}'$ of dyadic σ-algebras defined in Chapter 3.

THEOREM 5.1. *Let $2 < q \le \infty$. With equivalent norms,*

$$\text{BMO}_c^q(\mathbb{R}, \mathcal{M}) = \text{BMO}_c^{q,\mathcal{D}}(L^\infty(\mathbb{R}) \otimes \mathcal{M}) \cap \text{BMO}_c^{q,\mathcal{D}'}(L^\infty(\mathbb{R}) \otimes \mathcal{M});$$

$$\text{BMO}_r^q(\mathbb{R}, \mathcal{M}) = \text{BMO}_r^{q,\mathcal{D}}(L^\infty(\mathbb{R}) \otimes \mathcal{M}) \cap \text{BMO}_r^{q,\mathcal{D}'}(L^\infty(\mathbb{R}) \otimes \mathcal{M});$$

$$\text{BMO}_{cr}^q(\mathbb{R}, \mathcal{M}) = \text{BMO}_{cr}^{q,\mathcal{D}}(L^\infty(\mathbb{R}) \otimes \mathcal{M}) \cap \text{BMO}_{cr}^{q,\mathcal{D}'}(L^\infty(\mathbb{R}) \otimes \mathcal{M}).$$

Proof. From Proposition 3.1, $\forall t \in \mathbb{R}, h \in \mathbb{R}^+ \times \mathbb{R}^+$, there exist $k_{t,h}, N_h \in \mathbb{Z}$ such that $I_{h,t} := (t - h_1, t + h_2]$ is contained in $D_{N_h}^{k_{t,h}}$ or $D_{N_h}^{'k_{t,h}}$ and

$$|D_{N_h}^{k_{t,h}}| = |D_{N_h}^{'k_{t,h}}| \leq 6(h_1 + h_2).$$

If $I_{h,t} \subset D_{N_h}^{k_{t,h}}$, then

$$\begin{aligned}
\varphi_h^\#(t) &= \frac{1}{h_1 + h_2} \int_{t-h_1}^{t+h_2} |\varphi(x) - \varphi_{I_{h,t}}|^2 dx \\
&\leq \frac{4}{h_1 + h_2} \int_{t-h_1}^{t+h_2} |\varphi(x) - \varphi_{D_{N_h}^{k_{t,h}}}|^2 dx \\
&\leq \frac{24}{|D_{N_h}^{k_{t,h}}|} \int_{D_{N_h}^{k_{t,h}}} |\varphi(x) - \varphi_{D_{N_h}^{k_{t,h}}}|^2 dx \\
&\leq 24 \varphi_{\mathcal{D}_{N_h}}^\#(t).
\end{aligned}$$

Similarly, if $I_{h,t} \subset D_{N_h}^{'k_{t,h}}$, then

$$\varphi_h^\#(t) \leq 24 \varphi_{\mathcal{D}_{N_h}'}^\#(t).$$

Thus

$$\begin{aligned}
\|\varphi\|_{\mathrm{BMO}_c^q} &= \left\| \sup_{h \in \mathbb{R}^+ \times \mathbb{R}^+} |\varphi_h^\#| \right\|_{\frac{q}{2}}^{\frac{1}{2}} \\
&\leq \sqrt{24} \left\| \sup_n |(\varphi_{\mathcal{D}_n}^\# + \varphi_{\mathcal{D}_n'}^\#)| \right\|_{\frac{q}{2}}^{\frac{1}{2}} \\
&\leq 4\sqrt{3} \max(\|\varphi\|_{\mathrm{BMO}_c^{q,\mathcal{D}}}, \|\varphi\|_{\mathrm{BMO}_c^{q,\mathcal{D}'}}).
\end{aligned}$$

It is trivial that $\max(\|\varphi\|_{\mathrm{BMO}_c^{q,\mathcal{D}}}, \|\varphi\|_{\mathrm{BMO}_c^{q,\mathcal{D}'}}) \leq \|\varphi\|_{\mathrm{BMO}_c^q}$. Therefore

$$\mathrm{BMO}_c^q(\mathbb{R}, \mathcal{M}) = \mathrm{BMO}_c^{q,\mathcal{D}}(L^\infty(\mathbb{R}) \otimes \mathcal{M}) \cap \mathrm{BMO}_c^{q,\mathcal{D}'}(L^\infty(\mathbb{R}) \otimes \mathcal{M})$$

with equivalent norms. The two other equalities in the theorem are immediate consequences of this. ∎

2. The equivalence of $\mathcal{H}_{cr}^p(\mathbb{R}, \mathcal{M})$ and $L^p(L^\infty(\mathbb{R}) \otimes \mathcal{M})(1 < p < \infty)$

We denote the noncommutative martingale Hardy spaces defined in [28] and [16] with respect to $\mathcal{D}$ and $\mathcal{D}'$ by $\mathcal{H}_c^{p,\mathcal{D}}(L^\infty(\mathbb{R}) \otimes \mathcal{M}), \mathcal{H}_c^{p,\mathcal{D}'}(L^\infty(\mathbb{R}) \otimes \mathcal{M})$ etc.($1 \leq p < \infty$). Note that

$$\mathcal{H}_c^2(\mathbb{R}, \mathcal{M}) = \mathcal{H}_c^{2,\mathcal{D}}(L^\infty(\mathbb{R}) \otimes \mathcal{M}) = \mathcal{H}_c^{2,\mathcal{D}'}(L^\infty(\mathbb{R}) \otimes \mathcal{M}) = L^2(L^\infty(\mathbb{R}) \otimes \mathcal{M}).$$

By Theorems 4.4, 5.1 and the duality $(\mathcal{H}_c^{p,\mathcal{D}}(L^\infty(\mathbb{R}) \otimes \mathcal{M}))^* = \mathrm{BMO}_c^{q,\mathcal{D}}(L^\infty(\mathbb{R}) \otimes \mathcal{M})$ proved in [16], we get the following result.

COROLLARY 5.2. $\mathrm{BMO}_{cr}^q(\mathbb{R}, \mathcal{M}) = L^q(L^\infty(\mathbb{R}) \otimes \mathcal{M})$ *with equivalent norms for* $2 < q < \infty$.

Proof. From the inequalities (4.5) and (4.7) of [**16**] we have

$$\begin{aligned}
&\mathrm{BMO}_c^{q,\mathcal{D}}(L^\infty(\mathbb{R})\otimes\mathcal{M}) \cap \mathrm{BMO}_r^{q,\mathcal{D}}(L^\infty(\mathbb{R})\otimes\mathcal{M}) \\
&= L^q(L^\infty(\mathbb{R})\otimes\mathcal{M}) \\
&= \mathrm{BMO}_c^{q,\mathcal{D}'}(L^\infty(\mathbb{R})\otimes\mathcal{M}) \cap \mathrm{BMO}_r^{q,\mathcal{D}'}(L^\infty(\mathbb{R})\otimes\mathcal{M})
\end{aligned}$$

with equivalent norms. Therefore, by Theorem 5.1

$$\begin{aligned}
\mathrm{BMO}_{cr}^q(\mathbb{R},\mathcal{M}) &= \mathrm{BMO}_c^q(\mathbb{R},\mathcal{M}) \cap \mathrm{BMO}_r^q(\mathbb{R},\mathcal{M}) \\
&= \mathrm{BMO}_c^{q,\mathcal{D}}(L^\infty(\mathbb{R})\otimes\mathcal{M}) \cap \mathrm{BMO}_r^{q,\mathcal{D}}(L^\infty(\mathbb{R})\otimes\mathcal{M}) \\
&\quad \cap \mathrm{BMO}_c^{q,\mathcal{D}'}(L^\infty(\mathbb{R})\otimes\mathcal{M}) \cap \mathrm{BMO}_r^{q,\mathcal{D}'}(L^\infty(\mathbb{R})\otimes\mathcal{M}) \\
&= L^q(L^\infty(\mathbb{R})\otimes\mathcal{M}). \blacksquare
\end{aligned}$$

COROLLARY 5.3. *If* $1 \leq p < 2$, *then*

$$\begin{aligned}
\mathcal{H}_c^p(\mathbb{R},\mathcal{M}) &= \mathcal{H}_c^{p,\mathcal{D}}(L^\infty(\mathbb{R})\otimes\mathcal{M}) + \mathcal{H}_c^{p,\mathcal{D}'}(L^\infty(\mathbb{R})\otimes\mathcal{M}), \\
\mathcal{H}_r^p(\mathbb{R},\mathcal{M}) &= \mathcal{H}_r^{p,\mathcal{D}}(L^\infty(\mathbb{R})\otimes\mathcal{M}) + \mathcal{H}_r^{p,\mathcal{D}'}(L^\infty(\mathbb{R})\otimes\mathcal{M}), \\
\mathcal{H}_{cr}^p(\mathbb{R},\mathcal{M}) &= \mathcal{H}_{cr}^{p,\mathcal{D}}(L^\infty(\mathbb{R})\otimes\mathcal{M}) + \mathcal{H}_{cr}^{p,\mathcal{D}'}(L^\infty(\mathbb{R})\otimes\mathcal{M}).
\end{aligned}$$

If $p \geq 2$, *then*

$$\begin{aligned}
\mathcal{H}_c^p(\mathbb{R},\mathcal{M}) &= \mathcal{H}_c^{p,\mathcal{D}}(L^\infty(\mathbb{R})\otimes\mathcal{M}) \cap \mathcal{H}_c^{p,\mathcal{D}'}(L^\infty(\mathbb{R})\otimes\mathcal{M}), \\
\mathcal{H}_r^p(\mathbb{R},\mathcal{M}) &= \mathcal{H}_r^{p,\mathcal{D}}(L^\infty(\mathbb{R})\otimes\mathcal{M}) \cap \mathcal{H}_r^{p,\mathcal{D}'}(L^\infty(\mathbb{R})\otimes\mathcal{M}), \\
\mathcal{H}_{cr}^p(\mathbb{R},\mathcal{M}) &= \mathcal{H}_{cr}^{p,\mathcal{D}}(L^\infty(\mathbb{R})\otimes\mathcal{M}) \cap \mathcal{H}_{cr}^{p,\mathcal{D}'}(L^\infty(\mathbb{R})\otimes\mathcal{M}).
\end{aligned}$$

COROLLARY 5.4. $\mathcal{H}_{cr}^p(\mathbb{R},\mathcal{M}) = L^p(L^\infty(\mathbb{R})\otimes\mathcal{M})$ *with equivalent norms for all* $1 < p < \infty$.

Proof. Recall the result

$$\mathcal{H}_{cr}^{p,\mathcal{D}}(L^\infty(\mathbb{R})\otimes\mathcal{M}) = L^p(\mathbb{R},\mathcal{M}) = \mathcal{H}_{cr}^{p,\mathcal{D}'}(L^\infty(\mathbb{R})\otimes\mathcal{M})$$

proved in [**28**] and [**16**]. By Corollary 5.3, for $1 < p < 2$, we have

$$\begin{aligned}
\mathcal{H}_{cr}^p(\mathbb{R},\mathcal{M}) &= \mathcal{H}_c^p(\mathbb{R},\mathcal{M}) + \mathcal{H}_r^p(\mathbb{R},\mathcal{M}) \\
&= \mathcal{H}_c^{p,\mathcal{D}}(L^\infty(\mathbb{R})\otimes\mathcal{M}) + \mathcal{H}_c^{p,\mathcal{D}'}(L^\infty(\mathbb{R})\otimes\mathcal{M}) \\
&\quad + \mathcal{H}_r^{p,\mathcal{D}}(L^\infty(\mathbb{R})\otimes\mathcal{M}) + \mathcal{H}_r^{p,\mathcal{D}'}(L^\infty(\mathbb{R})\otimes\mathcal{M}) \\
&= \mathcal{H}_{cr}^{p,\mathcal{D}}(L^\infty(\mathbb{R})\otimes\mathcal{M}) + \mathcal{H}_{cr}^{p,\mathcal{D}'}(L^\infty(\mathbb{R})\otimes\mathcal{M}) \\
&= L^p(L^\infty(\mathbb{R})\otimes\mathcal{M})
\end{aligned}$$

and, for $2 \leq p < \infty$,

$$\begin{aligned}
\mathcal{H}_{cr}^p(\mathbb{R},\mathcal{M}) &= \mathcal{H}_c^p(\mathbb{R},\mathcal{M}) \cap \mathcal{H}_c^p(\mathbb{R},\mathcal{M}) \\
&= \mathcal{H}_c^{p,\mathcal{D}}(L^\infty(\mathbb{R})\otimes\mathcal{M}) \cap \mathcal{H}_c^{p,\mathcal{D}'}(L^\infty(\mathbb{R})\otimes\mathcal{M}) \\
&\quad \cap \mathcal{H}_r^{p,\mathcal{D}}(L^\infty(\mathbb{R})\otimes\mathcal{M}) \cap \mathcal{H}_r^{p,\mathcal{D}'}(L^\infty(\mathbb{R})\otimes\mathcal{M}) \\
&= \mathcal{H}_{cr}^{p,\mathcal{D}}(L^\infty(\mathbb{R})\otimes\mathcal{M}) \cap \mathcal{H}_{cr}^{p,\mathcal{D}'}(L^\infty(\mathbb{R})\otimes\mathcal{M}) \\
&= L^p(L^\infty(\mathbb{R})\otimes\mathcal{M}). \blacksquare
\end{aligned}$$

Remark. In [**18**] and [**19**], M. Junge, C. Le Merdy and Q. Xu studied the Littlewood-Paley theory for semigroups on noncommutative L^p-spaces. Among many results, they proved, in particular, that for many nice semigroups, the corresponding noncommutative Hardy spaces defined by the Littlewood-Paley g-function coincide with the underlying noncommutative L^p-spaces ($1 < p < \infty$). In their viewpoint, the semigroup in the context of our paper is the Poisson semigroup tensorized by the identity of $L^p(\mathcal{M})$. This semigroup satisfies all assumptions of [**19**]. Thus if we define our Hardy spaces $\mathcal{H}^p_{cr}(\mathbb{R}, \mathcal{M})$ by the g-function $G_c(f)$ and $G_r(f)$ (which is the same as that defined by $S_c(f)$ and $S_r(f)$ in virtue of Theorem 4.6), then Corollary 5.4 is a particular case of a general result from [**19**]. We should emphasize that the method in [**19**] is completely different from ours. It is based on the H^∞ functional calculus. It seems that the method in [**19**] does not permit to deal with the Lusin square functions $S_c(f)$ and $S_r(f)$.

CHAPTER 6

Interpolation

In this chapter, we consider interpolation for noncommutative Hardy spaces and BMO. The main results in this chapter are function space analogues of those in [23] for noncommutative martingales. On the other hand, they are also the extensions to the present noncommutative setting of the scalar results in [11]. Recall that the noncommutative L^p spaces associated with a semifinite von Neumann algebra form an interpolation scale with respect to both the complex and real interpolation methods. And, as the column (resp. row) subspaces of $L^p(\mathcal{M} \otimes B(L^2(\Omega)))$, the spaces $L^p(L^\infty(\mathbb{R}) \otimes \mathcal{M}, L_c^2(\widetilde{\Gamma}))$ form an interpolation scale also.

1. Complex interpolation

We first consider complex interpolation.

Let $\mathrm{BMO}_c^\mathcal{D}(L^\infty(\mathbb{R}) \otimes \mathcal{M})$ and $\mathcal{H}_c^{p,\mathcal{D}}(L^\infty(\mathbb{R}) \otimes \mathcal{M})$ (resp. $\mathrm{BMO}_c^{\mathcal{D}'}(L^\infty(\mathbb{R}) \otimes \mathcal{M})$ and $\mathcal{H}_c^{p,\mathcal{D}'}(L^\infty(\mathbb{R}) \otimes \mathcal{M})$) ($1 \le p < \infty$) be the noncommutative martingale BMO spaces and Hardy spaces defined in [16] with respect to the usual dyadic filtration $\mathcal{D}$ (resp. the dyadic filtration $\mathcal{D}'$) described in Chapter 3.

LEMMA 6.1. *For $1 < p < \infty$, we have*

(6.1) $(\mathrm{BMO}_c^\mathcal{D}(L^\infty(\mathbb{R}) \otimes \mathcal{M}), \mathcal{H}_c^{1,\mathcal{D}}(L^\infty(\mathbb{R}) \otimes \mathcal{M}))_{\frac{1}{p}} = \mathcal{H}_c^{p,\mathcal{D}}(L^\infty(\mathbb{R}) \otimes \mathcal{M}),$

(6.2) $(\mathrm{BMO}_r^\mathcal{D}(L^\infty(\mathbb{R}) \otimes \mathcal{M}), \mathcal{H}_r^{1,\mathcal{D}}(L^\infty(\mathbb{R}) \otimes \mathcal{M}))_{\frac{1}{p}} = \mathcal{H}_r^{p,\mathcal{D}}(L^\infty(\mathbb{R}) \otimes \mathcal{M}),$

(6.3) $\qquad\qquad\qquad (X, Y)_{\frac{1}{p}} = L^p(L^\infty(\mathbb{R}) \otimes \mathcal{M}).$

where $X = \mathrm{BMO}_{cr}^\mathcal{D}(L^\infty(\mathbb{R}) \otimes \mathcal{M})$ or $L^\infty(L^\infty(\mathbb{R}) \otimes \mathcal{M})$ and $Y = \mathcal{H}_{cr}^{1,\mathcal{D}}(L^\infty(\mathbb{R}) \otimes \mathcal{M})$ or $L^1(L^\infty(\mathbb{R}) \otimes \mathcal{M})$. Moreover, the same results hold for $\mathrm{BMO}_c^{\mathcal{D}'}(L^\infty(\mathbb{R}) \otimes \mathcal{M})$ and $\mathcal{H}_c^{p,\mathcal{D}'}(L^\infty(\mathbb{R}) \otimes \mathcal{M})$.

Proof. For each $k \in \mathbb{N}$ and each projection p of $\mathcal{M}$ with $\tau(p) < \infty$, denote by $\mathcal{H}_c^{q,\mathcal{D}}(L^\infty(-2^k, 2^k) \otimes p\mathcal{M}p)$ the subspace of $\mathcal{H}_c^{q,\mathcal{D}}(L^\infty(\mathbb{R}) \otimes \mathcal{M})$ consisting of elements supported on $(-2^k, 2^k)$ and with values in $p\mathcal{M}p$. By dualizing Theorem 3.1 of [23] we get, for $1 < r \le q < \infty$,

$$\left(\mathcal{H}_c^{1,\mathcal{D}}(L^\infty(-2^k, 2^k) \otimes p\mathcal{M}p), \mathcal{H}_c^{\frac{r}{r-1},\mathcal{D}}(L^\infty(-2^k, 2^k) \otimes p\mathcal{M}p)\right)_{\frac{r}{q}}$$
$$= \mathcal{H}_c^{\frac{q}{q-1},\mathcal{D}}(L^\infty(-2^k, 2^k) \otimes p\mathcal{M}p).$$

Note that the union of all these $\mathcal{H}_c^{r,\mathcal{D}}(L^\infty(-2^k, 2^k) \otimes p\mathcal{M}p)$ is dense in $\mathcal{H}_c^{r,\mathcal{D}}(L^\infty(\mathbb{R}) \otimes \mathcal{M})$. By approximation we get

(6.4) $(\mathcal{H}_c^{1,\mathcal{D}}(L^\infty(\mathbb{R}) \otimes \mathcal{M}), \mathcal{H}_c^{\frac{r}{r-1},\mathcal{D}}(L^\infty(\mathbb{R}) \otimes \mathcal{M}))_{\frac{r}{q}} = \mathcal{H}_c^{\frac{q}{q-1},\mathcal{D}}(L^\infty(\mathbb{R}) \otimes \mathcal{M})$

Dualizing (6.4) we have

(6.5) $\quad (\mathrm{BMO}_c^{\mathcal{D}}(L^\infty(\mathbb{R}) \otimes \mathcal{M}), \mathcal{H}_c^{r,\mathcal{D}}(L^\infty(\mathbb{R}) \otimes \mathcal{M}))_{\frac{r}{q}} = \mathcal{H}_c^{q,\mathcal{D}}(L^\infty(\mathbb{R}) \otimes \mathcal{M}).$

Combining (6.4) and (6.5) we get (6.1) by Wolff's interpolation theorem (see [**34**]). The equalities (6.2), (6.3) and the arguments for the dyadic filtration $\mathcal{D}'$ can be proved similarly.

THEOREM 6.2. *Let $1 < p < \infty$. Then with equivalent norms,*

(6.6) $\quad (\mathrm{BMO}_c(\mathbb{R},\mathcal{M}), \mathcal{H}_c^1(\mathbb{R},\mathcal{M}))_{\frac{1}{p}} = \mathcal{H}_c^p(\mathbb{R},\mathcal{M}),$

(6.7) $\quad (\mathrm{BMO}_r(\mathbb{R},\mathcal{M}), \mathcal{H}_r^1(\mathbb{R},\mathcal{M}))_{\frac{1}{p}} = \mathcal{H}_r^p(\mathbb{R},\mathcal{M}),$

(6.8) $\quad (X, Y)_{\frac{1}{p}} = L^p(L^\infty(\mathbb{R}) \otimes \mathcal{M}).$

where $X = \mathrm{BMO}_{cr}(\mathbb{R},\mathcal{M})$ or $L^\infty(L^\infty(\mathbb{R}) \otimes \mathcal{M})$ and $Y = \mathcal{H}_{cr}^1(\mathbb{R},\mathcal{M})$ or $L^1(L^\infty(\mathbb{R}) \otimes \mathcal{M})$.

Proof. Note that
$$\mathcal{H}_c^2(\mathbb{R},\mathcal{M}) = \mathcal{H}_c^{2,\mathcal{D}}(\mathbb{R},\mathcal{M}) = \mathcal{H}_c^{2,\mathcal{D}'}(\mathbb{R},\mathcal{M}).$$

Let $2 < q < \infty$. By Theorem 5.1 and Lemma 6.1 we have

$$(\mathrm{BMO}_c(\mathbb{R},\mathcal{M}), \mathcal{H}_c^2(\mathbb{R},\mathcal{M}))_{\frac{2}{q}}$$
$$= (\mathrm{BMO}_c^{\mathcal{D}}(L^\infty(\mathbb{R}) \otimes \mathcal{M}) \cap \mathrm{BMO}_c^{\mathcal{D}'}(L^\infty(\mathbb{R}) \otimes \mathcal{M}), \mathcal{H}_c^2(\mathbb{R},\mathcal{M}))_{\frac{2}{q}}$$
$$\subseteq (\mathrm{BMO}_c^{\mathcal{D}}(L^\infty(\mathbb{R}) \otimes \mathcal{M}), \mathcal{H}_c^2(\mathbb{R},\mathcal{M}))_{\frac{2}{q}} \cap (\mathrm{BMO}_c^{\mathcal{D}'}(L^\infty(\mathbb{R}) \otimes \mathcal{M}), \mathcal{H}_c^2(\mathbb{R},\mathcal{M}))_{\frac{2}{q}}$$
$$\subseteq \mathcal{H}_c^{q,\mathcal{D}}(L^\infty(\mathbb{R}) \otimes \mathcal{M}) \cap \mathcal{H}_c^{q,\mathcal{D}'}(L^\infty(\mathbb{R}) \otimes \mathcal{M})$$
$$= \mathcal{H}_c^q(\mathbb{R},\mathcal{M}).$$

Then by duality

(6.9) $\quad (\mathcal{H}_c^1(\mathbb{R},\mathcal{M}), \mathcal{H}_c^2(\mathbb{R},\mathcal{M}))_{\frac{2}{q}} \supseteq \mathcal{H}_c^{q'}(\mathbb{R},\mathcal{M}).$

The converse of (6.9) can be easily proved since the map Φ defined by $\Phi(f) = \nabla f(x+t,y)\chi_\Gamma(x,y)$ is isometric from $\mathcal{H}_c^{q'}(\mathbb{R},\mathcal{M})$ to $L^{q'}(L^\infty(\mathbb{R}) \otimes \mathcal{M}, L_c^2(\widetilde{\Gamma}))$ for $q \geq 1$. Thus we have

(6.10) $\quad (\mathcal{H}_c^1(\mathbb{R},\mathcal{M}), \mathcal{H}_c^2(\mathbb{R},\mathcal{M}))_{\frac{2}{q}} = \mathcal{H}_c^{q'}(\mathbb{R},\mathcal{M}).$

Dualizing this equality once more, we get

(6.11) $\quad (\mathrm{BMO}_c(\mathbb{R},\mathcal{M}), \mathcal{H}_c^2(\mathbb{R},\mathcal{M}))_{\frac{2}{q}} = \mathcal{H}_c^q(\mathbb{R},\mathcal{M}).$

Note that by Proposition 2.1 and Theorem 4.8, $\mathcal{H}_c^q$ is complemented in $L^q(L^\infty(\mathbb{R}) \otimes \mathcal{M}, L_c^2(\widetilde{\Gamma}))(1 < q < \infty)$ via the embedding Φ. Hence, from the interpolation result (1.3) we have

(6.12) $\quad (\mathcal{H}_c^q(\mathbb{R},\mathcal{M}), \mathcal{H}_c^{q'}(\mathbb{R},\mathcal{M}))_{\frac{1}{2}} = \mathcal{H}_c^2(\mathbb{R},\mathcal{M})$

Combining (6.10), (6.11) and (6.12) we get (6.6) by Wolff's interpolation theorem (see [**34**]). (6.7) can be proved similarly. For (6.8), by Lemma 6.1 and Theorem

5.1,
$$\begin{aligned}&(\mathrm{BMO}_{cr}(\mathbb{R},\mathcal{M}), L^1(L^\infty(\mathbb{R})\otimes\mathcal{M}))_{\frac{1}{p}}\\ =\;&(\mathrm{BMO}_{cr}^{\mathcal{D}}(L^\infty(\mathbb{R})\otimes\mathcal{M})\cap\mathrm{BMO}_{cr}^{\mathcal{D}'}(L^\infty(\mathbb{R})\otimes\mathcal{M}), L^1(L^\infty(\mathbb{R})\otimes\mathcal{M}))_{\frac{1}{p}}\\ \subseteq\;&(\mathrm{BMO}_{cr}^{\mathcal{D}}(L^\infty(\mathbb{R})\otimes\mathcal{M}), L^1(L^\infty(\mathbb{R})\otimes\mathcal{M}))_{\frac{1}{p}}\\ &\cap(\mathrm{BMO}_{cr}^{\mathcal{D}'}(L^\infty(\mathbb{R})\otimes\mathcal{M}), L^1(L^\infty(\mathbb{R})\otimes\mathcal{M}))_{\frac{1}{p}}\\ =\;&L^p(L^\infty(\mathbb{R})\otimes\mathcal{M})\end{aligned}$$

On the other hand, since $\mathrm{BMO}_{cr}(\mathbb{R},\mathcal{M})\supset L^\infty(L^\infty(\mathbb{R})\otimes\mathcal{M})$,

$$\begin{aligned}&(\mathrm{BMO}_{cr}(\mathbb{R},\mathcal{M}), L^1(L^\infty(\mathbb{R})\otimes\mathcal{M}))_{\frac{1}{p}}\\ \supseteq\;&(L^\infty(L^\infty(\mathbb{R})\otimes\mathcal{M}), L^1(L^\infty(\mathbb{R})\otimes\mathcal{M}))_{\frac{1}{p}}\\ =\;&L^p(L^\infty(\mathbb{R})\otimes\mathcal{M}).\end{aligned}$$

Therefore,
$$(\mathrm{BMO}_{cr}(\mathbb{R},\mathcal{M}), L^1(L^\infty(\mathbb{R})\otimes\mathcal{M}))_{\frac{1}{p}} = L^p(L^\infty(\mathbb{R})\otimes\mathcal{M}).$$

By duality we have
$$(L^\infty(L^\infty(\mathbb{R})\otimes\mathcal{M}), \mathcal{H}_{cr}^1(\mathbb{R},\mathcal{M}))_{\frac{1}{p}} = L^p(L^\infty(\mathbb{R})\otimes\mathcal{M}).$$

Finally,
$$\begin{aligned}(L^\infty(L^\infty(\mathbb{R})\otimes\mathcal{M}), \mathcal{H}_{cr}^1(\mathbb{R},\mathcal{M}))_{\frac{1}{p}} &\subseteq (\mathrm{BMO}_{cr}(\mathbb{R},\mathcal{M}), \mathcal{H}_{cr}^1(\mathbb{R},\mathcal{M}))_{\frac{1}{p}}\\ &\subseteq (\mathrm{BMO}_{cr}(\mathbb{R},\mathcal{M}), L^1(L^\infty(\mathbb{R})\otimes\mathcal{M}))_{\frac{1}{p}}.\end{aligned}$$

Hence
$$(\mathrm{BMO}_{cr}(\mathbb{R},\mathcal{M}), \mathcal{H}_{cr}^1(\mathbb{R},\mathcal{M}))_{\frac{1}{p}} = L^p(L^\infty(\mathbb{R})\otimes\mathcal{M}).$$

Thus we have obtained all equalities in the theorem. ∎

Remark. We know little about $(\mathrm{BMO}_c(\mathbb{R},\mathcal{M}), L^1(L^\infty(\mathbb{R})\otimes\mathcal{M})_{\frac{1}{p}}$ even for $p=2$.

2. Real interpolation

The following theorem concerns real interpolation.

THEOREM 6.3. *Let $1\leq p<\infty$. Then with equivalent norms,*

(6.13) $$(X,Y)_{\frac{1}{p},p} = L^p(L^\infty(\mathbb{R})\otimes\mathcal{M}).$$

where $X=\mathrm{BMO}_{cr}(\mathbb{R},\mathcal{M})$ or $L^\infty(L^\infty(\mathbb{R})\otimes\mathcal{M})$ and $Y=\mathcal{H}_{cr}^1(\mathbb{R},\mathcal{M})$ or $L^1(L^\infty(\mathbb{R})\otimes\mathcal{M})$.

Proof. By Theorem 4.3 of [**23**] and Theorem 5.1 we have (using the same argument as above for the complex method)

$$(\mathrm{BMO}_{cr}(\mathbb{R},\mathcal{M}), L^1(L^\infty(\mathbb{R})\otimes\mathcal{M}))_{\frac{1}{p},p} \subseteq L^p(L^\infty(\mathbb{R})\otimes\mathcal{M}).$$

On the other hand, for $1<p<\infty$,

$$\begin{aligned}(\mathrm{BMO}_{cr}(\mathbb{R},\mathcal{M}), L^1(L^\infty(\mathbb{R})\otimes\mathcal{M}))_{\frac{1}{p},p} &\supseteq (L^\infty(L^\infty(\mathbb{R})\otimes\mathcal{M}), L^1(L^\infty(\mathbb{R})\otimes\mathcal{M}))_{\frac{1}{p},p}\\ &= L^p(L^\infty(\mathbb{R})\otimes\mathcal{M}).\end{aligned}$$

Therefore
$$(\mathrm{BMO}_{cr}(\mathbb{R},\mathcal{M}), L^1(L^\infty(\mathbb{R})\otimes\mathcal{M}))_{\frac{1}{p},p} = L^p(L^\infty(\mathbb{R})\otimes\mathcal{M}), \quad 1<p<\infty.$$

By duality we have
$$(L^\infty(L^\infty(\mathbb{R})\otimes\mathcal{M}), \mathcal{H}^1_{cr}(\mathbb{R},\mathcal{M}))_{\frac{1}{p},p} = L^p(L^\infty(\mathbb{R})\otimes\mathcal{M}), \quad 1<p<\infty.$$

Noting again that
$$\begin{aligned}(L^\infty(L^\infty(\mathbb{R})\otimes\mathcal{M}), \mathcal{H}^1_{cr}(\mathbb{R},\mathcal{M}))_{\frac{1}{p},p} &\subseteq (\mathrm{BMO}_{cr}(\mathbb{R},\mathcal{M}), \mathcal{H}^1_{cr}(\mathbb{R},\mathcal{M}))_{\frac{1}{p},p}\\ &\subseteq (\mathrm{BMO}_{cr}(\mathbb{R},\mathcal{M}), L^1(L^\infty(\mathbb{R})\otimes\mathcal{M}))_{\frac{1}{p},p},\end{aligned}$$

we conclude
$$\mathrm{BMO}_{cr}(\mathbb{R},\mathcal{M}), \mathcal{H}^1_{cr}(\mathbb{R},\mathcal{M}))_{\frac{1}{p},p} = L^p(L^\infty(\mathbb{R})\otimes\mathcal{M})), \quad 1<p<\infty. \blacksquare$$

3. Fourier multipliers

We close this chapter with a result on Fourier multipliers. Recall that $H^1(\mathbb{R})$ denotes the classical Hardy space on $\mathbb{R}$. We will also need $H^1(\mathbb{R}, H)$, the H^1 on $\mathbb{R}$ with values in a Hilbert space H. Recall that we say a bounded map $M: H^1(\mathbb{R}) \to H^1(\mathbb{R})$ is a Fourier multiplier if there exists a function $m \in L^\infty(\mathbb{R})$ such that

$$\widehat{Mf} = m\widehat{f}, \quad \forall f \in H^1(\mathbb{R})$$

where $\widehat{f}$ is the Fourier transform of f.

THEOREM 6.4. *Let M be a Fourier multiplier of the classical Hardy space $H^1(\mathbb{R})$. Then M extends in a natural way to a bounded map on $\mathrm{BMO}_c(\mathbb{R}, \mathcal{M})$ and $\mathcal{H}^p_c(\mathbb{R}, \mathcal{M})$ for all $1 \leq p < \infty$ and*

$$(6.14) \quad \|M: \mathrm{BMO}_c(\mathbb{R},\mathcal{M}) \to \mathrm{BMO}_c(\mathbb{R},\mathcal{M})\| \leq c \|M: H^1(\mathbb{R}) \to H^1(\mathbb{R})\|,$$
$$(6.15) \quad \|M: \mathcal{H}^p_c(\mathbb{R},\mathcal{M}) \to \mathcal{H}^p_c(\mathbb{R},\mathcal{M})\| \leq c \|M: H^1(\mathbb{R}) \to H^1(\mathbb{R})\|.$$

Similar results also hold for $\mathrm{BMO}_r(\mathbb{R},\mathcal{M}), \mathrm{BMO}_{cr}(\mathbb{R},\mathcal{M}), \mathcal{H}^p_c(\mathbb{R},\mathcal{M})$ *and* $\mathcal{H}^p_{cr}(\mathbb{R},\mathcal{M})$.

Proof. Assume $\|M: H^1(\mathbb{R}) \to H^1(\mathbb{R})\| = 1$. Let H be the Hilbert space on which $\mathcal{M}$ acts. We start by showing the (well known) fact that M is bounded on $H^1(\mathbb{R}, H)$. Denote by R the Hilbert transform. Recall that $\|f\|_{H^1(\mathbb{R},H)} \simeq \|f\|_{L^1(\mathbb{R},H)} + \|Rf\|_{L^1(\mathbb{R},H)}$ for every $f \in H^1(\mathbb{R}, H)$. Denote by $\{e_\lambda\}_{\lambda \in \Lambda}$ the orthogonal normalized basis of H. Then $f = (f_\lambda)_{\lambda \in \Lambda}$ with $f_\lambda = \langle e_\lambda, f \rangle e_\lambda$. Note that if $f \in H^1(\mathbb{R}, H)$ then at most countably many f_λ's are non zero. Let $\varepsilon = (\varepsilon_n)_{n \in \mathbb{N}}$ be a sequence of independent random variables on some probability space (Ω, P) such that $P(\varepsilon_n = 1) = P(\varepsilon_n = -1) = \frac{1}{2}, \forall n \in \mathbb{N}$. Notice that $MR = RM$. Let $f \in H^1(\mathbb{R}, H)$. Let $\{\lambda_n : n \in \mathbb{N}\}$ be an enumeration of the λ's such that $f_\lambda \neq 0$. Then by Khintchine's inequality,

$$\begin{aligned}
\|Mf\|_{H^1(\mathbb{R},H)} &\simeq \int_{\mathbb{R}}((\sum_{n\in\mathbb{N}}|Mf_{\lambda_n}|^2)^{\frac{1}{2}} + (\sum_{n\in\mathbb{N}}|RMf_{\lambda_n}|^2)^{\frac{1}{2}})dt \\
&\simeq \int_{\mathbb{R}}\int_{\Omega}|\sum_{n\in\mathbb{N}}\varepsilon_n Mf_{\lambda_n}|dP(\varepsilon)dt + \int_{\mathbb{R}}\int_{\Omega}|\sum_{n\in\mathbb{N}}\varepsilon_n MRf_{\lambda_n}|dP(\varepsilon)dt \\
&\simeq \int_{\Omega}\left\|M(\sum_{n\in\mathbb{N}}\varepsilon_n f_{\lambda_n})\right\|_{H^1(\mathbb{R},H)} dP(\varepsilon) \\
&\leq c\int_{\Omega}\left\|\sum_{n\in\mathbb{N}}\varepsilon_n f_{\lambda_n}\right\|_{H^1(\mathbb{R},H)} dP(\varepsilon) \\
&\leq c\|f\|_{H^1(\mathbb{R},H)}
\end{aligned}$$

Therefore, as announced
$$\|M : H^1(\mathbb{R},H) \to H^1(\mathbb{R},H)\| \leq c_1.$$

Then by transposition
$$\|M : \mathrm{BMO}(\mathbb{R},H) \to \mathrm{BMO}(\mathbb{R},H)\| \leq c_2;$$

whence, in virtue of (1.16),
$$\|M : \mathrm{BMO}_c(\mathbb{R},\mathcal{M}) \to \mathrm{BMO}_c(\mathbb{R},\mathcal{M})\| \leq c_2.$$

Thus by duality
$$\|M : \mathcal{H}_c^1(\mathbb{R},\mathcal{M}) \to \mathcal{H}_c^1(\mathbb{R},\mathcal{M})\| \leq c_3.$$

Then by Theorem 6.1 we have
$$\|M : \mathcal{H}_c^p(\mathbb{R},\mathcal{M}) \to \mathcal{H}_c^p(\mathbb{R},\mathcal{M})\| \leq c_4.$$

Hence we have obtained the assertion concerning the column spaces. The other assertions are immediate consequences of this one. ∎

Remark. Very recently, Junge and Musat got a John-Nirenberg theorem for BMO spaces of noncommutative martingales (see [15]). By using Proposition 3.1 and the dyadic trick of this article, they got a John-Nirenberg theorem for noncommutative BMO spaces discussed here (which can also be proved as a consequence of the interpolation results established in this chapter). Unlike the classical case, the value of

$$\sup_{I\subset\mathbb{R}}\left\|\left(\frac{1}{|I|}\int_I |\varphi - \varphi_I|^p d\mu\right)^{\frac{1}{p}}\right\|_{\mathcal{M}}$$

for different p, $0 < p < \infty$ are no longer equivalent to each other. In fact, if $\mathcal{M} = M_n$ the algebra of n by n matrices, it can be proved that the best constant c_n such that

$$\sup_{I\subset\mathbb{R}}\left\|\frac{1}{|I|}\int_I |\varphi - \varphi_I|^2 d\mu\right\|_{M_n}^{\frac{1}{2}} \leq c_n \sup_{I\subset\mathbb{R}}\left\|\frac{1}{|I|}\int_I |\varphi - \varphi_I| d\mu\right\|_{M_n},$$

holds for all $\varphi \in \mathrm{BMO}_c(\mathbb{R}, M_n)$ will be at least $c\log n$ as $n \to \infty$. And the corresponding constant for M_n valued martingales could be $cn^{\frac{1}{2}}$ if no additional assumption on the related filtration. What remains true is the equivalence of

$$\sup_{I\subset\mathbb{R}}\sup_{\tau|a|^p\leq 1}|I|^{-\frac{1}{p}}\|(f-f_I)a\chi_I\|_{L^p(\mathbb{R},\mathcal{M})} + \sup_{I\subset\mathbb{R}}\sup_{\tau|a|^p\leq 1}|I|^{-\frac{1}{p}}\|a\chi_I(f-f_I)\|_{L^p(\mathbb{R},\mathcal{M})}$$

for different $p, 2 \leq p < \infty$ (see Theorem 1.2 of [**15**]) and the equivalence of

$$\sup_{\text{cube } I \subset \mathbb{R}} \sup_{\tau |a|^p \leq 1,} \{|I|^{-\frac{1}{p}} \|(f - f_I)a\chi_I\|_{\mathcal{H}_c^p(\mathbb{R},\mathcal{M})}\}$$

for different $p, 2 \leq p < \infty$. See [**15**], [**22**] for more information on this.

Acknowledgments. Most of the work carried out in this paper was done under the direction of Quanhua Xu. The author is greatly indebted to Q. Xu for having suggested to him the subject of this paper, for many helpful discussion and very careful reading of this paper. He has paid a lot of attention to all parts of this paper and made many essential modifications to its original version. The author is also very grateful to his current advisor Gilles Pisier for his very careful reading of the manuscript and very useful comments which led to many corrections and improvements. Finally, the author would like to thank the referee for his very careful reading of the manuscript and very useful comments.

Bibliography

[1] J. S. Bendat, S., Sherman, Monotone and convex operator functions. Trans. Amer. Math. Soc. 79 (1955), 58–71.
[2] Bhatia, Rajendra, Matrix analysis. Graduate Texts in Mathematics, 169. Springer-Verlag, New York, 1997.
[3] I. Cuculescu, Martingales on von Neumann algebras. J. Multiv. Anal., 1 (1971), 17-27.
[4] T. Fack, H. Kosaki, Generalized s-Numbers of τ-Measurable Operators, Pacific. J. Math.,123 (1986), 269-300.
[5] C. Fefferman, E.M. Stein, H^p spaces of several variables, Acta Math., 129 (1972), 137-193.
[6] J. B. Garnett, Bounded Analytic Functions, Pure and Applied Mathematics, 96. Academic press, Inc., New York-London, 1981.
[7] A. Garsia, Martingale Inequalities, Sem. Notes on Recent Progress, Benjamin, 1973.
[8] T. A. Gillespie, S. Pott, S. Treil, A. Volberg, Logarithmic Growth for Martingale transform, Journal of London Mathematical Society(2), 64 (2001), no. 3, 624-636.
[9] F. Hansen, An Operator Inequality, Math. Ann., 246 (1979/80), no. 3, 249–250.
[10] C. S. Herz, H_p−Spaces of Martingales, $1 < p \leq 1$, Zeit, Wahrscheinlichkeits theorie, 28(1974), 189-265.
[11] S. Janson, P. W. Jones, Interpolation between H^p spaces: the complex method, Journal of Functional Analysis, 48 (1982), no. 1, 58-80.
[12] R. Jajte, Strong Limit Theorems in noncommutative Probability, Lecture Notes in Math., 1110, Springer-Verlag, Berlin, 1985.
[13] R. Jajte, Strong Limit Theorems in noncommutative L_2-Spaces, Lecture Notes in Mathematics, 1477. Springer-Verlag, Berlin, 1991.
[14] M. Junge, Doob's Inequality for noncommutative Martingales, J. Reine Angew. Math. 549 (2002), 149-190.
[15] M. Junge, M. Musat, A noncommutative version of the John-Nirenberg theorem, Transactions of AMS, to appear.
[16] M. Junge, Q. Xu, Noncommutative Burkholder/Rosenthal Inequalities, Ann. Prob. 31 (2003), no. 2, 948-995.
[17] M. Junge, Q. Xu, Noncommutative Maximal Ergodic Theorems, preprint.
[18] M. Junge, C. Le Merdy, Q. Xu, H^∞ Functional Calculus and Square Functions on noncommutative L^p-spaces, C.R. Acad. Sci. Paris, 337 (2003), 93-98.
[19] M. Junge, C. Le Merdy, Q. Xu, H^∞ Functional Calculus and Square Functions on noncommutative L^p-spaces, Preprint.
[20] P. Koosis, Introduction to H_p Spaces, Cambridge Tracts in Mathematics, 115, Cambridge, 1998.
[21] T. Mei, BMO is the intersection of two translates of dyadic BMO, C.R. Acad. Sci. Paris, 336 (2003), 1003-1006.
[22] T. Mei, Remark on BMO Spaces in the noncommutative Setting , Preprint.
[23] M. Musat, Interpolation Between noncommutative BMO and noncommutative L_p-spaces, Journal of Functional Analysis, 202 (2003), no. 1, 195-225.
[24] F. Nazarov, G. Pisier, S. Treil, A. Volberg, Sharp Estimates in Vector Carleson Imbedding Theorem and for Vector Paraproducts. J. Reine Angew. Math. 542 (2002), 147-171.
[25] S. Petermichl. Dyadic Shifts and a Logarithmic Estimate for Hankel Operators with Matrix Symbol. C.R. Acad. Sci. Paris, 330 (2000), no. 6,455-460.
[26] G. Pisier, Notes on Banach Space Valued H^p-Spaces, preprint.
[27] G. Pisier, noncommutative Vector Valued L_p-Spaces and Completely p-Summing Maps, Soc. Math. France. Astérisque (1998) 237.

[28] G. Pisier, Q. Xu, noncommutative Martingale Inequalities, Comm. Math. Phys. 189 (1997), 667-698.
[29] S. Pott, M. Smith, Vector Paraproducts. and Hankel operators of Schatten Class via p-John-Nirenberg theorem, J. Funct. Anal. 217(2004), no. 1, 38-78.
[30] E. Ricard, Décomposition de H^1, Multiplicateurs de Schur et Espaces d'Operateurs, Thèse de Doctorat de l'Université Paris VI. 2001.
[31] N. Randrianantoanina, noncommutative martingale transform, Journal of Functional Analysis, 194 (2002), no. 1, 181-212.
[32] E. M. Stein, Harmonic Analysis, Princeton Univ. Press, Princeton, New Jersey, 1993.
[33] E. M. Stein, Harmonic Analysis, Princeton Univ. Press, Princeton, New Jersey, 1970.
[34] T. Wolff, A Note on Interpolation Spaces, Harmonic Analysis, Lec. Notes in Math., 1568, Spring-Verlag, Berlin-Heidelberg-New York, 1994.

Math. Dept., Texas A&M Univ., College Station, TX, 77843, U. S. A. , and Math. Dept., WuHan Univ., 430072, P.R. China; email: tmei@math.tamu.edu

Editorial Information

To be published in the *Memoirs*, a paper must be correct, new, nontrivial, and significant. Further, it must be well written and of interest to a substantial number of mathematicians. Piecemeal results, such as an inconclusive step toward an unproved major theorem or a minor variation on a known result, are in general not acceptable for publication.

Papers appearing in *Memoirs* are generally at least 80 and not more than 200 published pages in length. Papers less than 80 or more than 200 published pages require the approval of the Managing Editor of the Transactions/Memoirs Editorial Board.

As of February 28, 2007, the backlog for this journal was approximately 15 volumes. This estimate is the result of dividing the number of manuscripts for this journal in the Providence office that have not yet gone to the printer on the above date by the average number of monographs per volume over the previous twelve months, reduced by the number of volumes published in four months (the time necessary for preparing a volume for the printer). (There are 6 volumes per year, each usually containing at least 4 numbers.)

A Consent to Publish and Copyright Agreement is required before a paper will be published in the *Memoirs*. After a paper is accepted for publication, the Providence office will send a Consent to Publish and Copyright Agreement to all authors of the paper. By submitting a paper to the *Memoirs*, authors certify that the results have not been submitted to nor are they under consideration for publication by another journal, conference proceedings, or similar publication.

Information for Authors

Memoirs are printed from camera copy fully prepared by the author. This means that the finished book will look exactly like the copy submitted.

Initial submission. The AMS uses Centralized Manuscript Processing for initial submissions. Authors should submit a PDF file using the Initial Manuscript Submission form found at www.ams.org/cgi-bin/peertrack/submission.pl, or send one copy of the manuscript to the following address: Centralized Manuscript Processing, MEMOIRS OF THE AMS, 201 Charles Street, Providence, RI 02904-2294 USA. If a paper copy is being forwarded to the AMS, indicate that it is for it Memoirs and include the name of the corresponding author, contact information such as email address or mailing address, and the name of an appropriate Editor to review the paper (see the list of Editors below).

The paper must contain a *descriptive title* and an *abstract* that summarizes the article in language suitable for workers in the general field (algebra, analysis, etc.). The *descriptive title* should be short, but informative; useless or vague phrases such as "some remarks about" or "concerning" should be avoided. The *abstract* should be at least one complete sentence, and at most 300 words. Included with the footnotes to the paper should be the 2000 *Mathematics Subject Classification* representing the primary and secondary subjects of the article. The classifications are accessible from www.ams.org/msc/. The list of classifications is also available in print starting with the 1999 annual index of *Mathematical Reviews*. The Mathematics Subject Classification footnote may be followed by a list of *key words and phrases* describing the subject matter of the article and taken from it. Journal abbreviations used in bibliographies are listed in the latest *Mathematical Reviews* annual index. The series abbreviations are also accessible from www.ams.org/publications/. To help in preparing and verifying references, the AMS offers MR Lookup, a Reference Tool for Linking, at www.ams.org/mrlookup/.

Electronically prepared manuscripts. The AMS encourages electronically prepared manuscripts, with a strong preference for $\mathcal{AMS}$-LaTeX. To this end, the Society has prepared $\mathcal{AMS}$-LaTeX author packages for each AMS publication. Author packages include instructions for preparing electronic manuscripts, samples, and a style file that generates the particular design specifications of that publication series. Though $\mathcal{AMS}$-LaTeX is the highly preferred format of TeX, author packages are also available in $\mathcal{AMS}$-TeX.

Authors may retrieve an author package from the AMS website starting from www.ams.org/tex/ or via FTP to ftp.ams.org (login as anonymous, enter username as password, and type cd pub/author-info). The *AMS Author Handbook* and the *Instruction Manual* are available in PDF format following the author packages link from www.ams.org/tex/. The author package can also be obtained free of charge by sending email to tech-support@ams.org (Internet) or from the Publication Division, American Mathematical Society, 201 Charles St., Providence, RI 02904-2294, USA. When requesting an author package, please specify $\mathcal{AMS}$-LaTeX or $\mathcal{AMS}$-TeX and the publication in which your paper will appear. Please be sure to include your complete mailing address.

After acceptance. The final version of the electronic file should be sent to the Providence office (this includes any TeX source file, any graphics files, and the DVI or PostScript file) immediately after the paper has been accepted for publication.

Before sending the source file, be sure you have proofread your paper carefully. The files you send must be the EXACT files used to generate the proof copy that was accepted for publication. For all publications, authors are required to send a printed copy of their paper, which exactly matches the copy approved for publication, along with any graphics that will appear in the paper.

Accepted electronically prepared files can be submitted via the web at www.ams.org/submit-book-journal/, sent via FTP, or sent on CD-Rom or diskette to the Electronic Prepress Department, American Mathematical Society, 201 Charles Street, Providence, RI 02904-2294 USA. TeX source files, DVI files, and PostScript files can be transferred over the Internet by FTP to the Internet node ftp.ams.org (130.44.1.100). When sending a manuscript electronically via CD-Rom or diskette, please be sure to include a message identifying the paper as a Memoir.

Electronically prepared manuscripts can also be sent via email to pub-submit@ams.org (Internet). In order to send files via email, they must be encoded properly. (DVI files are binary and PostScript files tend to be very large.)

Electronic graphics. Comprehensive instructions on preparing graphics are available at www.ams.org/jourhtml/. A few of the major requirements are given here.

Submit files for graphics as EPS (Encapsulated PostScript) files. This includes graphics originated via a graphics application as well as scanned photographs or other computer-generated images. If this is not possible, TIFF files are acceptable as long as they can be opened in Adobe Photoshop or Illustrator. No matter what method was used to produce the graphic, it is necessary to provide a paper copy to the AMS.

Authors using graphics packages for the creation of electronic art should also avoid the use of any lines thinner than 0.5 points in width. Many graphics packages allow the user to specify a "hairline" for a very thin line. Hairlines often look acceptable when proofed on a typical laser printer. However, when produced on a high-resolution laser imagesetter, hairlines become nearly invisible and will be lost entirely in the final printing process.

Screens should be set to values between 15% and 85%. Screens which fall outside of this range are too light or too dark to print correctly. Variations of screens within a graphic should be no less than 10%.

Inquiries. Any inquiries concerning a paper that has been accepted for publication should be sent to memo-query@ams.org or directly to the Electronic Prepress Department, American Mathematical Society, 201 Charles St., Providence, RI 02904-2294 USA.

Editors

This journal is designed particularly for long research papers, normally at least 80 pages in length, and groups of cognate papers in pure and applied mathematics. Papers intended for publication in the *Memoirs* should be addressed to one of the following editors. The AMS uses Centralized Manuscript Processing for initial submissions to AMS journals. Authors should follow instructions listed on the Initial Submission page found at www.ams.org/memo/memosubmit.html.

Algebra to ALEXANDER KLESHCHEV, Department of Mathematics, University of Oregon, Eugene, OR 97403-1222; email: ams@noether.uoregon.edu

Algebra and its application to MINA TEICHER, Emmy Noether Research Institute for Mathematics, Bar-Ilan University, Ramat-Gan 52900, Israel; email: teicher@macs.biu.ac.il

Algebraic geometry to DAN ABRAMOVICH, Department of Mathematics, Brown University, Box 1917, Providence, RI 02912; email: amsedit@math.brown.edu

Algebraic number theory to V. KUMAR MURTY, Department of Mathematics, University of Toronto, 100 St. George Street, Toronto, ON M5S 1A1, Canada; email: murty@math.toronto.edu

Algebraic topology to ALEJANDRO ADEM, Department of Mathematics, University of British Columbia, Room 121, 1984 Mathematics Road, Vancouver, British Columbia, Canada V6T 1Z2; email: adem@math.ubc.ca

Combinatorics to JOHN R. STEMBRIDGE, Department of Mathematics, University of Michigan, Ann Arbor, Michigan 48109-1109; email: FRS@umich.edu

Complex analysis and harmonic analysis to ALEXANDER NAGEL, Department of Mathematics, University of Wisconsin, 480 Lincoln Drive, Madison, WI 53706-1313; email: nagel@math.wisc.edu

Differential geometry and global analysis to LISA C. JEFFREY, Department of Mathematics, University of Toronto, 100 St. George St., Toronto, ON Canada M5S 3G3; email: jeffrey@math.toronto.edu

Dynamical systems and ergodic theory to AMIE WILKINSON, Department of Mathematics, Northwestern University, 2033 Sheridan Road, Evanston, IL 60208-2730; email: transactions@math.northwestern.edu

Functional analysis and operator algebras to DIMITRI SHLYAKHTENKO, Department of Mathematics, University of California, Los Angeles, CA 90095; email: shlyakht@math.ucla.edu

Geometric analysis to WILLIAM P. MINICOZZI II, Department of Mathematics, Johns Hopkins University, 3400 N. Charles St., Baltimore, MD 21218; email: trans@math.jhu.edu

Geometric analysis to MLADEN BESTVINA, Department of Mathematics, University of Utah, 155 South 1400 East, JWB 233, Salt Lake City, Utah 84112-0090; email: bestvina@math.utah.edu

Harmonic analysis, representation theory, and Lie theory to ROBERT J. STANTON, Department of Mathematics, The Ohio State University, 231 West 18th Avenue, Columbus, OH 43210-1174; email: stanton@math.ohio-state.edu

Logic to STEFFEN LEMPP, Department of Mathematics, University of Wisconsin, 480 Lincoln Drive, Madison, Wisconsin 53706-1388; email: lempp@math.wisc.edu

Partial differential equations to GUSTAVO PONCE, Department of Mathematics, South Hall, Room 6607, University of California, Santa Barbara, CA 93106; email: ponce@math.ucsb.edu

Partial differential equations and dynamical systems to PETER POLACIK, School of Mathematics, University of Minnesota, Minneapolis, MN 55455; email: polacik@math.umn.edu

Probability and statistics to KRZYSZTOF BURDZY, Department of Mathematics, University of Washington, Box 354350, Seattle, Washington 98195-4350; email: burdzy@math.washington.edu

Real analysis and partial differential equations to DANIEL TATARU, Department of Mathematics, University of California, Berkeley, Berkeley, CA 94720; email: tataru@math.berkeley.edu

All other communications to the editors should be addressed to the Managing Editor, ROBERT GURALNICK, Department of Mathematics, University of Southern California, Los Angeles, CA 90089-1113; email: guralnic@math.usc.edu.

Titles in This Series

883 **Apostolos Beligiannis and Idun Reiten,** Homological and homotopical aspects of torsion theories, 2007

882 **Lars Inge Hedberg and Yuri Netrusov,** An axiomatic approach to function spaces, spectral synthesis, and Luzin approximation, 2007

881 **Tao Mei,** Operator valued Hardy spaces, 2007

880 **Bruce C. Berndt, Geumlan Choi, Youn-Seo Choi, Heekyoung Hahn, Boon Pin Yeap, Ae Ja Yee, Hamza Yesilyurt, and Jinhee Yi,** Ramanujan's forty identities for the Rogers-Ramanujan functions, 2007

879 **O. García-Prada, P. B. Gothen, and V. Muñoz,** Betti numbers of the moduli space of rank 3 parabolic Higgs bundles, 2007

878 **Alessandra Celletti and Luigi Chierchia,** KAM stability and celestial mechanics, 2007

877 **María J. Carro, José A. Raposo, and Javier Soria,** Recent developments in the theory of Lorentz spaces and weighted inequalities, 2007

876 **Gabriel Debs and Jean Saint Raymond,** Borel liftings of Borel sets: Some decidable and undecidable statements, 2007

875 **C. Krattenthaler and T. Rivoal,** Hypergéométrie et fonction zêta de Riemann, 2007

874 **Sonia Natale,** Semisolvability of semisimple Hopf algebras of low dimension, 2007

873 **A. J. Duncan,** Exponential genus problems in one-relator products of groups, 2007

872 **Anthony V. Geramita, Tadahito Harima, Juan C. Migliore, and Yong Su Shin,** The Hilbert function of a level algebra, 2007

871 **Pascal Auscher,** On necessary and sufficient conditions for L^p-estimates of Riesz transforms associated to elliptic operators on $\mathbb{R}^n$ and related estimates, 2007

870 **Takuro Mochizuki,** Asymptotic behaviour of tame harmonic bundles and an application to pure twistor D-modules, Part 2, 2007

869 **Takuro Mochizuki,** Asymptotic behaviour of tame harmonic bundles and an application to pure twistor D-modules, Part 1, 2007

868 **Gelu Popescu,** Entropy and multivariable interpolation, 2006

867 **Vilmos Totik,** Metric properties of harmonic measures, 2006

866 **William Craig,** Semigroups underlying first-order logic, 2006

865 **Nathanial P. Brown,** Invariant means and finite representation theory of $C*$-algebras, 2006

864 **John M. Lee,** Fredholm operators and Einstein metrics on conformally compact manifolds, 2006

863 **M. Lübke and A. Teleman,** The Universal Kobayashi-Hitchin correspondence on Hermitian manifolds, 2006

862 **Alberto Canonaco,** The Beilinson complex and canonical rings of irregular surfaces, 2006

861 **Leon A. Takhtajan and Lee-Peng Teo,** Weil-Petersson metric on the universal Teichmüller space, 2006

860 **Thomas M. Fiore,** Pseudo limits, biadjoints and pseudo algebras: Categorical foundations of conformal field theory, 2006

859 **N. Arcozzi, R. Rochberg, and E. Sawyer,** Carleson measures and interpolating sequences for Besov spaces on complex balls, 2006

858 **Enrico Valdinoci, Berardino Sciunzi, and Vasile Ovidiu Savin,** Flat level set regularity of p-Laplace phase transitions, 2006

857 **Donatella Danielli, Nocola Garofalo, and Duy-Minh Nhieu,** Non-doubling Ahlfors measures, perimeter measures, and the characterization of the trace spaces of Sobolev functions in Carnot-Carathéodory spaces, 2006

856 **Vladimir Bolotnikov and Harry Dym,** On boundary interpolation for matrix valued Schur functions, 2006

TITLES IN THIS SERIES

855 **Yevgenia Kashina, Yorck Sommerhäuser, and Yongchang Zhu,** On higher Frobenius-Schur indicators, 2006

854 **Noam Greenberg,** The role of true finiteness in the admissible recursively enumerable degrees, 2006

853 **Joachim Krieger,** Stability of spherically symmetric wave maps, 2006

852 **Viorel Barbu, Irena Lasiecka, and Roberto Triggiani,** Tangential boundary stabilization of Navier-Stokes equations, 2006

851 **Jie Wu,** On maps from loop suspensions to loop spaces and the shuffle relations on the Cohen groups, 2006

850 **Siegfried Echterhoff, S. Kaliszewski, John Quigg, and Iain Raeburn,** A categorical approach to imprimitivity theorems for C^*-dynamical systems, 2006

849 **Katsuhiko Kuribayashi, Mamoru Mimura, and Tetsu Nishimoto,** Twisted tensor products related to the cohomology of the classifying spaces of loop groups, 2006

848 **Bob Oliver,** Equivalences of classifying spaces completed at the prime two, 2006

847 **Eric T. Sawyer and Richard L. Wheeden,** Hölder continuity of weak solutions to subelliptic equations with rough coefficients, 2006

846 **Victor Beresnevich, Detta Dickinson, and Sanju Velani,** Measure theoretic laws for lim–sup sets, 2006

845 **Ehud Friedgut, Vojtech Rödl, Andrzej Ruciński, and Prasad V. Tetali,** A Sharp threshold for random graphs with a monochromatic triangle in every edge coloring, 2006

844 **Amadeu Delshams, Rafael de la Llave, and Tere M. Seara,** A geometric mechanism for diffusion in Hamiltonian systems overcoming the large gap problem: Heuristics and rigorous verification on a model, 2006

843 **Denis V. Osin,** Relatively hyperbolic groups: Intrinsic geometry, algebraic properties, and algorithmic problems, 2006

842 **David P. Blecher and Vrej Zarikian,** The calculus of one-sided M-ideals and multipliers in operator spaces, 2006

841 **Enrique Artal Bartolo, Pierrette Cassou-Noguès, Ignacio Luengo, and Alejandro Melle Hernández,** Quasi-ordinary power series and their zeta functions, 2005

840 **Sławomir Kołodziej,** The complex Monge-Ampère equation and pluripotential theory, 2005

839 **Mihai Ciucu,** A random tiling model for two dimensional electrostatics, 2005

838 **V. Jurdjevic,** Integrable Hamiltonian systems on complex Lie groups, 2005

837 **Joseph A. Ball and Victor Vinnikov,** Lax-Phillips scattering and conservative linear systems: A Cuntz-algebra multidimensional setting, 2005

836 **H. G. Dales and A. T.-M. Lau,** The second duals of Beurling algebras, 2005

835 **Kiyoshi Igusa,** Higher complex torsion and the framing principle, 2005

834 **Keníchi Ohshika,** Kleinian groups which are limits of geometrically finite groups, 2005

833 **Greg Hjorth and Alexander S. Kechris,** Rigidity theorems for actions of product groups and countable Borel equivalence relations, 2005

832 **Lee Klingler and Lawrence S. Levy,** Representation type of commutative Noetherian rings III: Global wildness and tameness, 2005

831 **K. R. Goodearl and F. Wehrung,** The complete dimension theory of partially ordered systems with equivalence and orthogonality, 2005

For a complete list of titles in this series, visit the
AMS Bookstore at **www.ams.org/bookstore/**.